HAND-CRANKED PHONOGRAPHS

It All Started With Edison....

Thomas Alva Edison
1847 - 1931

HAND-CRANKED PHONOGRAPHS

It All Started With Edison....

An Introduction to Vintage Talking Machines, Records and More!

BY
NEIL MAKEN

With forewords by
Oliver Berliner
and
George Tselos

PROMAR PUBLISHING   **HUNTINGTON BEACH, CA**

Maken, Neil
   Hand-Cranked Phonographs – It All Started With Edison....

   1. Antiques
   2. Collectibles
   3. Sound Recording
   4. History of Sound Recording

   LCCC #93-084079

   First Edition – 1993
        Second printing – 1993
        Third printing – 1994
        Fourth printing – 1995
        Fifth printing – 1998 with updated bibliography

   ISBN 0-9640687-1-0 Paperback
   ISBN 0-9640687-0-2 Hardbound

Cover photograph: Thomas Alva Edison after a non-stop 72 hour marathon working on the phonograph (see page 3)

Frontpiece Illustration: Painting of Thomas Alva Edison dated November 26, 1926

Additional copies of this book or a free catalog of other books on hand-cranked phonographs and recordings may be ordered directly from:
Yesterday Once Again, PO Box 6773, Huntington Beach, CA 92615 USA

"THE MUSIC GOES 'ROUND AND AROUND ... AND IT COMES OUT HERE."

"Red" Hodgson / (Select Music, 1935)

"...THERE AIN'T NO RULES AROUND HERE! WE'RE TRYING TO ACCOMPLISH SOMEP'N."

Thomas Alva Edison

HAND-CRANKED PHONOGRAPHS

It All Started With Edison....

TABLE OF CONTENTS

THIS BOOK IS DEDICATED WITH LOVE TO MY THREE WOMEN:

TO CAROLE, for knowing better than to be underfoot.

TO TRACEY, for not being around enough to be underfoot.

TO CLEO, for making up for the others....

Forewords

Rather than another romance of the recording industry, this book presents hard facts about the talking machine, the records and some of the companies that made them. The reader must look elsewhere for historical details, and this book tells where to look - just as it explains common terms used in the industry and by collectors - and provides practical help on restoring and evaluating the machines that have stood the test of time, some more than a century old and still working.

Emile Berliner didn't set out to invent or to even introduce disk records. His intention was to create a method of mass producing unlimited copies of a recording - be it cylinder or disk - from a single original record. Today's multi-billion dollar recording industry is vocal testimony to the fact that he succeeded. And many characteristics of today's electronic disks contain myriad characteristics of my grandfather's primitive acoustical creation.

This book may serve to turn a cursory interest into an intense one. And many rarely known facts are revealed to whet one's appetite and show why things were done the way they were. In this respect there's a wealth of information here for the novice, the serious collector and for that special person who's fortunate enough to own a special piece of history in what introduced low cost professional entertainment into every home, taught the plowboy to whistle grand opera and gave us "the music you want, when you want it".

Oliver Berliner, 1993

It was not until I assumed the position of Archivist at the Edison National Historic Site in 1989 that I began to appreciate the amount of interest that exists around the country in collecting early phonographs and sound recordings. The staff of the Edison Archives receives hundreds of questions each year from people anxious to obtain information about Edison phonograph machines as well as cylinder and disc records which they have bought or are thinking of buying, have inherited, or just found in their grandparents' attic. Neil Maken, a long time friend of the Edison National Historic Site, has put together a book which will undoubtedly be of great use to beginning collectors in this field. Its clear explanations and illustrations may reduce to some degree the number of basic questions which are directed our way by the public, but will inevitably result in these same phonograph fans coming back again in the future with more sophisticated inquiries.

George Tselos, 1993

Acknowledgements -

This is perhaps, the most difficult portion of the book to write. There are so many people I wish to thank for their invaluable help over the years, but I could never even attempt to do so without unintentionally slighting someone. I have made many friends, both far and near, through these marvelous talking machines. And from each of them I have gained a little more knowledge. They have all been very helpful and open in sharing their expertise and answering numerous questions. I hope to repay them in some small part with this book by answering the questions of the next generations of collectors.

I can only thank a handful personally. Of them, and topping the list, is my wife, Carole. With the patience of a saint and the wisdom of Solomon she has tolerated, advised and fully supported my hobby, and, as it turned into a business, she again was there lending her fullest support. Her home was over-run by my machines. She listened patiently when she really didn't want to know, and allowed me free rein. As the business grew, she became even more supportive through my seven-day work week which often began at five in the morning and ended after ten at night. And she is still there, and her enthusiasm for me and my obsession has not diminished. Carole, **thank you!**

Others whom I wish to thank personally include my good friend Oliver Berliner, who virtually singlehandedly chose to restore his grandfather's rightful due in the development of the Gramophone. He researched, traveled, wrote numerous letters, and over the years uncovered a great deal about Emile Berliner which he generously shared with the world.

George Tselos, the archivist at the Edison National Historic Site in West Orange, NJ has been extremely helpful in researching the great Thomas Edison. He put up with untold phone calls and visits, openly shared his findings, mailed numerous photocopies and never once complained. And a special thanks too, to Leah Burt, former curator of the Edison Home, *Glenmont*, and her archival digging.

And to George Frow: George - thank you! You are a wealth of information, a tireless researcher, a stickler for detail and one of the most open and sharing people I know. Your books, your labors of love, have enriched the phonograph-collecting world immeasurably. Thank you. And an extra-special thanks to you George for your kind permission to reprint so many photographs from your own archives.

Ken King: thanks for your technical support and advice, and for your friendship. Kent Flygare for all off your help and encouragement - you are sorely missed.

And finally, but certainly not least, to my daughter, Tracey, who is the love of my life and my inspiration. For her complete support, cooperation and friendship; for sharing my hobby so that she might share my time: TJ, I love you.

Preface

This volume is not intended to be a complete text on the invention and development of the phonograph. Neither is it meant to list every manufacturer nor to illustrate every phonograph. And it is not intended to be a guide to the current values of these vintage talking machines.

What the intent is though, is to give the phonograph owner and the new collector an overview of the history of these wonderful machines, and to help identify different models and manufacturers so that an owner can properly appreciate and enjoy his phonograph. Over the years each manufacturer incorporated many changes into his talking machine. Add-on accessories were also made available so that the phonograph owner during the first fifty years or so of the phonograph could update his machine and make it compatible with the improvements being made within the industry.

My objective is to highlight the major manufacturers and their products, and to offer a guide to the identification of groups and models of phonographs.

The time span of this book is from the invention of the phonograph by Thomas Edison in 1877 to approximately 1929. 1929 was the year the Edison Company ceased production of phonographs and records. It was the year in which the Radio Corporation of America purchased the Victor Talking Machine Company (becoming RCA-Victor), and it can be viewed as an arbitrary end to the acoustic recording era.

We will examine the major phonograph manufacturers and their most popular products, but we will also look at a few of the rare and exotic specimens as well. Then we will view a few phonographs by some of the lesser known but still very interesting and collectible companies.

I sincerely hope that this volume whets your appetite for more and detailed information. There is a wealth of scholarly knowledge available for those seeking it. Rather than compete with these fine existing works, I hope to be the stimulus to make you want to know more about this exciting and often profitable hobby of collecting vintage talking machines.

Neil Maken
1993

HAND-CRANKED PHONOGRAPHS

It All Started With Edison....

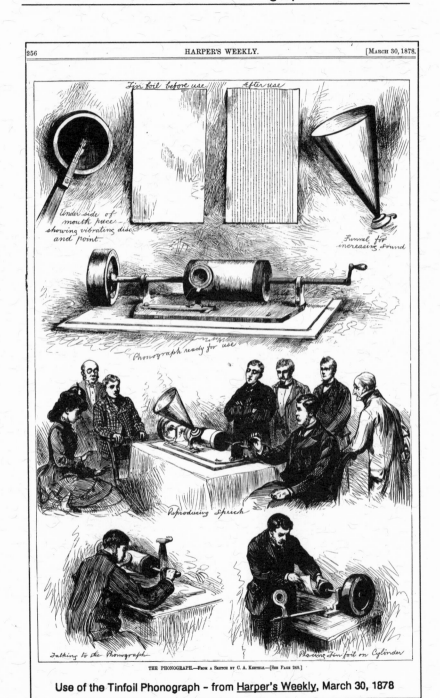

Use of the Tinfoil Phonograph – from <u>Harper's Weekly</u>, March 30, 1878

IT ALL STARTED WITH EDISON...

...well actually, it started <u>before</u> Edison. As far back as 1856, Edouard
Leon Scott de Martinville produced a machine which would mark sound
waves onto a smoke blackened drum. He named his machine
<u>phonautograph</u>, from the Greek, meaning "sound signature". (Scott's
cylinder recorder was just that: it could make recordings but they were
unplayable.) In early 1877, Charles Cros presented the idea of using
tracings to make a playable record. In late 1877 Thomas Edison would
create a playable recording, and dropping the letters a-u-t from Scott's
word, he called his invention a <u>phonograph</u>. It was Edison who actually
produced a machine capable of both recording and playing back voices!

Edison's '*Phonograph*' was a totally new concept for people of a less
sophisticated time to understand. They knew that a music box was
capable of playing a melody from metal pins on either a disc or cylinder,
but to actually hear a voice speaking or a band playing from a machine
was more than many could comprehend.

But it did record and play back voices and music. Edison's very first
phonograph used a large drum, around which was wrapped a layer of
tinfoil (a soft material of tin and lead, similar in appearance to aluminum
foil). One stylus attached to a diaphragm would record sounds onto the
tinfoil; the second stylus would later trace the same patterns and convert
them into audible sound. Within a very short period of time several
companies were manufacturing phonographs. Edison's time was needed
in other areas, primarily working on developing a suitable incandescent
light and an electrical distribution system, and the phonograph was put
onto a back burner for future development.

In the early 1880's, Alexander Graham Bell invited his cousin, Chichester
Bell, and a family friend, Charles Sumner Tainter, to investigate the
many mysteries of sound. They took the Edison phonograph and
replaced the tinfoil with a wax cylinder. The wax was more conducive to
recording sound than was the tinfoil, and the finished 'record' was more
permanent than the tinfoil. Whereas Edison named his talking machine
the **PHONOGRAPH**, Bell and Tainter switched syllables around and
called theirs the **GRAPHOPHONE**. The competition of Bell and
Tainter was a challenge to Edison, and after years of not having worked
on his phonograph, Edison was back in the race for improvements to
make his invention superior to all challengers. With his associates,
Edison spent a 72 hour virtually non-stop period on the phonograph and
emerged with the *PERFECTED* phonograph, an electric machine
designed for business use as a dictating machine. Owners of the
phonograph took advantage of the machine's versatility and started

The GRAMMY AWARD, the highest honor of the recording industry. This award was presented posthumously to "Eldridge R. Johnson, Industry Pioneer... 1985"
Courtesy Delaware State Museums, Dover, DE

making home recordings of songs, stories and music. So began the multi-billion dollar industry of sound recording that we know today.

The earliest commercial cylinder records were of a brown-wax composition, and were extremely fragile. Duplication methods were primitive: pantographing, used as early as c.1896 (a mechanical method of copying a duplicate from an original) was the only way of copying from a master record. Each recording was, in fact, an original. During a typical recording session, perhaps five machines were set up to record the performer. If one hundred finished records were needed, it might require twenty 'takes' by the performer. An error or an outside noise could easily ruin five records before they were completed.

In 1887, ten years after Edison's initial invention of the machine that talked, Emile Berliner, a German immigrant working in Washington D.C., solved the problem of duplication from a master record. He developed a *disk* or platter shaped record. Using a system of coating a zinc disk with wax and inscribing the wax with a stylus attached to a diaphragm, the sound waves would cause the stylus to cut through the wax coating. The disk was then immersed in acid; the acid would etch the zinc where the wax had been removed. From this zinc 'master' a 'stamper' consisting of a 'negative' of the master would be made. From the stamper a great number of records could be made, quickly and inexpensively.

To play these new disk records, Berliner developed a talking machine which he called a **GRAMOPHONE**. A stylus attached to a diaphragm would 'feel' the physical impressions of sound in the grooves and convert them to audible sound. Berliner's earliest Gramophones were hand-cranked affairs: the turntable was attached through a belt drive to a crank. As long as the crank was turned the record would revolve. But turn the crank too slowly and the speed and pitch of the record would diminish; turn too quickly and the sound would climb in pitch and speed.

Recording & Details of the Gramophone – <u>Scientific American</u>, May 16, 1896

Berliner had a machinist named Eldridge R. Johnson of Camden, New Jersey build a Gramophone which incorporated a spring in the motor. The speed of the turntable was now independent of the speed of cranking (a governor assembly regulated the speed), and the sound was considerably more consistent.

The talking machine industry in the mid-1890's consisted of two formats of records: disks and cylinders, and three major companies: Edison's Phonograph, Bell & Tainter's Graphophone and Berliner & Johnson's Gramophone. Looking back now, the pre-1895 talking machine industry was primitive. The recording methods were in their infantile years, and the talking machines themselves were unwieldy affairs. Power sources for the talking machines were being examined and tried. The direct drive system was the easiest: the crank was connected to the turntable (either directly or through a belt drive). Electricity was available from wet cell batteries, and phonograph motors were developed to use this power source. Later, line current electricity was applied.

Besides their cylinder graphophone, the Bell & Tainter company offered a talking machine mounted on a stand with a treadle to provide a source of power. Edison had a battery operated phonograph. He also experimented with a water-powered phonograph and a finger-powered phonograph as well as his own treadle-powered model.

In 1890 Edison even introduced the first novelty phonograph: a miniaturized hand-cranked phonograph in a doll's body. In theory, if not in practice, it was a wonderful idea. Now a doll could actually talk! But the brown-wax records were too delicate to survive the use by a child with her doll, and this endeavor was short-lived. Berliner also introduced a talking doll in 1890, but this enterprise did not prove successful. Several years later, a French phonograph manufacturer, M. Lioret, inserted a spring wound miniaturized phonograph into the body of the renowned Jumeau doll. It was years later and this phonograph was considerably more advanced in technology than Edison's. It used a jeweled stylus rather than the metal stylus of Edison. The records were of a celluloid material and much less subject to wear and breakage than the early brown wax. In the 'teens and 20's, several phonograph talking dolls were introduced which used a celluloid record.

The Edison Phonograph Doll, the first talking doll, circa 1890.

The novelty of the phonograph encouraged its purchase, and it was not uncommon for a family to be invited to a 'phonograph' party. Neighbors would be invited to the 'One Home' in the area which had a phonograph to hear the newest songs and stories on records. Before long these neighbors would purchase their own phonograph, and it became a part of the normal home furnishings.

For a detailed history of the talking machine, please refer to the Welsh & Burt book From Tinfoil to Stereo.

A variety of types of phonographs. From left to right: outside horn cylinder phonograph, concealed horn cylinder phonograph, outside horn disk phonograph, floor model concealed horn disk phonograph, tabletop concealed horn phonograph and floor model Edison Diamond Disc phonograph.

AN INTRODUCTION TO PHONOGRAPHS: A BASIC PRIMER

Rather than jump immediately into a discussion of phonographs and manufacturers, it would be wise to take a moment to do an over-view of the range of hand-cranked phonographs. If you have been collecting for some time, this is basic and second nature to you; however if you are new to the fascinating world of vintage phonographs, this is a good starting place. It will answer many of the questions that you have, and hopefully, create many more.

There are essentially two formats of phonographs: cylinder and disk. These can be further divided into outside or exposed horn machines and inside or concealed horn machines. As a rule of thumb, the outside horn phonographs were manufactured earlier than the inside horn models (although outside horn machines were manufactured well into the first quarter of the twentieth century). A further division will be found in the cabinetry: table model or floor model phonographs.

A.   OUTSIDE HORN CYLINDER PHONOGRAPHS: manufactured from approximately 1898 to 1912 by Edison and Columbia. Cabinet styles, detailing, features and horns varied depending on manufacturer and price, but all were basically similar. Essentially all were table-top models, but many were sold with matching or coordinating under-cabinets, often containing storage areas for the cylinder records.

B.   INSIDE (CONCEALED) HORN CYLINDER PHONOGRAPHS: introduced by Edison, he called these phonographs *AMBEROLAS* for the new four-minute Amberol records, and slightly later for the special Blue Amberol records that they played. Amberolas were introduced about 1909 and were manufactured well into the 1920's. They were available both in table-top models and in floor models.

C.   OUTSIDE HORN DISC PHONOGRAPHS: available from the late 1890's through the mid-1920's, disk phonographs were manufactured by a number of companies such as Columbia, Victor and Duplex and had a bracket and cradle on the front of the phonograph. These were known as *front-mounted horn* phonographs.

Later machines had a support arm bolted to the cabinet back which supported the horn and the tone-arm assembly. These are known as *back-mount* or *rear-mount* horns. The outside horn disk phonographs were generally all table-top models, although some had matching bases available.

D.   INSIDE HORN (CONCEALED HORN) PHONOGRAPHS: in an effort to make the phonograph look less like an appliance and more like a piece of furniture, the entire motor, tone arm and horn were placed inside a cabinet. Often there was a lid which, when closed, concealed the phonograph completely. The horn, rather than being exposed above the turntable, was designed to curve down into the cabinet. The sound emerged through a set of doors, louvers or a decorative grill.

1.   Floor model versions of the phonograph generally stood approximately four feet high (and were about 24" square). Possibly hundreds of models and manufacturers were available. With phonographs the 'hot-ticket' in the 1920's, everyone wanted some of the action. Many smaller companies came and went very quickly. Besides containing the phonograph motor, tone arm and horn, these also provided for storage of the disk records, either on shelves or in specially designed vertical slots. These phonographs (from about 1906 through the 1930's) were generally available in either oak or mahogany veneer. From about 1920 on, a modified upright cabinet called a console or low-boy was introduced. These cabinets also contained the phonograph and storage areas, but stood only about 30" high, and were approximately four feet long.

2.   Tabletop models of the inside horn disk phonograph were generally budget versions, with simple cabinets containing only the phonograph but no storage area. They were designed to sit on a table, shelf or an optional storage cabinet. They were available from about 1910 through the mid-1920's.

E.   EDISON DIAMOND DISC PHONOGRAPHS:
      special mention should be made of Edison's entry into
      the inside-horn phonograph business. Edison believed
      in <u>vertical recording</u> (discussed later in the book) and
      his disc phonograph was special in that it only played
      special vertically recorded Edison Disc records. The
      playback facilities were not compatible with the regular
      78 RPM phonographs of the day. Edison produced his
      disc phonographs from about 1913 through the 1920's.

FOR A ROUGH TIMETABLE IN DATING A PHONOGRAPH,
REFER TO THE CHART BELOW:

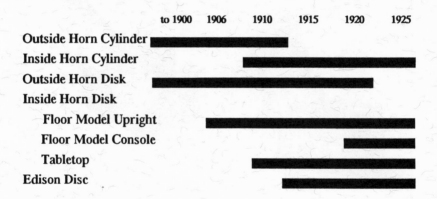

	to 1900	1906	1910	1915	1920	1925
Outside Horn Cylinder
Inside Horn Cylinder
Outside Horn Disk
Inside Horn Disk
Floor Model Upright
Floor Model Console
Tabletop
Edison Disc

Edison offered a variety of cylinder phonographs at
costs from $7.50 and up.

EDISON'S CYLINDER PHONOGRAPHS

Thomas Edison was a man of the people. After his initial resistance to using the phonograph for anything other than a business machine, he sought to provide the public with a phonograph that was reasonably priced, well constructed, reliable and easy to use.

As improvements were made, they were incorporated into Edison's phonographs. As a result, new models were introduced regularly, each incorporating the improvements of the previous model.

It is with model designation that confusion abounds. Edison manufactured phonographs to fit into a range of price points: with increasing price came better quality, but he constantly strove to incorporate improvements within each price range.

The *GEM* was initially designed to retail at $7.50; the *STANDARD* at $20, the *HOME* at $40, the *TRIUMPH* at $50; the *OPERA* at $90; and the luxury *IDELIA* at $125. Over the years prices increased for these phonographs, and in 1909, the *FIRESIDE* was introduced at $22 (to fill the price point vacated by the *STANDARD* which was now selling for $30).

Within each of these models however, was an entire series of improved models designated by a letter. The letter designation (ranging in most cases from 'A' up) indicated the major improvements made within that type of phonograph over the previous letter designations. (e.g. a Model D STANDARD was a two and four minute combination phonograph, whereas its predecessors, the Models A, B and C were of the two-minute-only type.) But improvements were constantly being made on the phonographs, and within any letter designation one might find two or three variations. Most of the early models did not indicate the letter designation on the machine. It is only by examining details on each machine that one can properly identify the letter model. (The error is often made by the new collector of mistakenly assigning the model letter found on the reproducer to the phonograph. We will discuss this further in the section on reproducers.) We highly recommend the book the *Edison Cylinder Phonograph Companion* by George Frow for further details on the model variations and dating of Edison cylinder phonographs.

To further complicate matters, Edison sold parts for the owner (or dealer) to update existing phonographs, These parts would fit onto or replace parts on the older machine and make them compatible with the latest technology. It is quite common to now find a Model A or B *HOME* or *STANDARD* upgraded to play both two and four minute records.

The Edison OPERA, a true luxury phonograph for the discriminating buyer.

The AMBEROLA 1A was Edison's first concealed-horn phonograph

By 1909 Edison had recognized that the success of Victor's new Victrola with a concealed horn was the trend of the future. To answer the affluent public's demand for a piece of furniture, Edison introduced the *AMBEROLA*, a luxurious cabinet in mahogany or oak enclosing the phonograph mechanism and horn within a fine piece of furniture. Originally designed to play both the two-minute and the four-minute wax Amberol cylinder records, the AMBEROLA soon progressed to playing only four-minute records, and in 1912 was further modified to play only the new celluloid *Blue Amberol* records introduced by Edison. These records, with their hard surface, were best played with a diamond stylus. As with many of his other phonographs, Edison developed an entire line of AMBEROLAs at price points to satisfy most economic levels. Today the *AMBEROLA 30*, introduced in 1915 at $30, is quite often found having survived the years in good condition.

Today, the Edison cylinder phonograph models most often found are the *STANDARD*, the *HOME* and the *AMBEROLA 30*. This is a direct reflection of the early popularity of these phonographs and of the durability which allowed them to withstand the passage of decades with minimal deterioration.

The AMBEROLA 30 was a durable phonograph in a simple oak cabinet.

The *HOME* was introduced in 1896. It utilized a single mainspring, and the feed screw was actually an extension of the mandrel shaft. Physically, it was larger than the *STANDARD* (introduced just a couple of years later). The early *HOME* was known as the 'banner-lid' model. It had suitcase-type clips for the lid, and the lid itself was emblazoned with a decal proclaiming "*Edison Home Phonograph*".

The *STANDARD* was introduced in 1898, and the earliest models were housed in a simple oak cabinet with a square-cornered lid. The mechanism was compact, and the feed screw was set behind the record mandrel and parallel to it. The single-mainspring motor was concealed within the cabinet bottom, and power was transferred to the upper mechanism by way of a leather drive belt. The cylinder record would fit

The earliest STANDARD of 1898 was encased in a square oak cabinet and lid, but the mechanism varied very little from later models. Later models had a domed lid.

onto a tapered mandrel, and the carriage assembly would traverse the mandrel. The carriage assembly consisted of the carriage housing, a reproducer, (and on the earliest models, a shaver mechanism) and a feed-nut or half-nut to move the carriage from left to right. When set in the 'play' position, the feed-nut at the back of the carriage assembly would engage a feed screw to provide the left-right movement, and the stylus of the reproducer would float in the grooves of the record and convert the impressions in the record to audible sound. At the extreme right end of the record mandrel was a swing gate which supported the mandrel and positioned it properly for play. In 1901 the square lid of the *STANDARD* was replaced with a curved top lid, and owners of the earlier cabinets could update their phonographs merely by purchasing a new style cabinet.

The first of the new style cabinets had a banner decal on the front declaring it to be an "*Edison Standard Phonograph*" or an "*Edison Home Phonograph*". These early phonographs were designed for home recording, and came with a recorder for making records. The early *STANDARD*s and *HOME*s played <u>only</u> two-minute cylinder records. Initially they were equipped with an 'Automatic' style reproducer, but this soon gave way to the much more popular **MODEL 'C'** reproducer. The 'C' reproducer would play <u>only</u> two-minute cylinders.

The *STANDARD* and *HOME* progressed through a series of improvements including elimination of the shaving device, elimination of the end-gate and replacement of the banner decal with the single word *Edison*, in signature style, on the cabinet front.

In 1908, Edison introduced a cylinder record of four-minute duration. To play these records, a new gearing system and a smaller stylus were required. The *STANDARD* Model D was the first to incorporate both two and four-minute gearing from the factory. By merely pulling or pushing a knob at the extreme left hand side of the upper works one could shift from two-minute to four-minute. The *HOME* Model D was also a two and four minute combination, using a slightly more complex gearing system. The shifting mechanism for the *HOME* looked like a second pulley, adjacent to and immediately to the right of the belt pulley. By sliding the shifting 'pulley' to the right, the four minute gearing was engaged; by sliding to the left, the machine was put into the two-minute position.

The four-minute records were physically the same size as the two-minute records, but they had twice as many grooves on their surface. The grooves and the walls between them were much finer than the earlier two-minute records. A new reproducer was introduced to play these four-minute records. The *MODEL* 'H' reproducer was specifically designed to play four-minute cylinders. A *MODEL* 'K' reproducer was introduced to play both two and four-minute cylinders without changing reproducers. The 'K' had two stylus bars, one two-minute and one four-minute, and they could be easily rotated from one to the other rather than exchanging the entire reproducer.

The popularity of the four-minute records created a demand among owners of older phonographs for both two and four-minute capability. Edison answered these demands with a factory-built conversion kit. Essentially a bolt-on affair, it replaced the single two-minute gear with a combination two-minute and four-minute gear.

The conversion unit for the *STANDARD* required removing the cast iron gear cover on the left side of the upper works, replacing the single two-minute gear on the mandrel shaft, and bolting a shifting mechanism to the left side frame of the upper works. The entire conversion was then shrouded under a new gear cover. A two-minute *STANDARD* converted to both two and four minute can be easily recognized by the gear cover. The replacement gear cover slopes at an angle from the front to the rear. The factory equipped two and four minute models used a gear cover which was horizontal.

The gear cover for the STANDARD 2/4 minute conversion unit slopes from front to back (below).

The factory equipped 2/4 minute STANDARD has a horizontal gear cover (above).

The 2-and-4 minute conversion unit for the STANDARD was a simple bolt-on affair.

If the gear cover is missing, examining the gears on the extreme left end of the mandrel shaft will show either a single gear (2-minute) or a double gear (2 and 4-minute model).

The single gear (left) is 2 minute only.

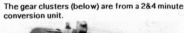

The gear clusters (below) are from a 2&4 minute conversion unit.

The *HOME* conversion unit required removing the existing belt pulley and replacing it with a special 'double' pulley assembly which actually enclosed the shifting mechanism. A bolt-on lever allowed the user to select either the two-minute or four-minute mode.

The Edison *TRIUMPH*, although much larger in motor and cabinet, was very similar to the *HOME*, utilizing the same feed screw configuration and the same conversion system to two and four-minute.

The Edison *GEM* was at first a two-minute only phonograph (in the black metal housing), and later a two and four-minute combination (in the maroon housing) and later still, four-minute only, also in a maroon housing. The four-minute only *GEM* phonographs are quite collectible.

In 1912 Edison introduced his newest development: a Blue Amberol cylinder record of four-minute duration. The cylinder records were blue in color, and unlike the earlier black and brown cylinders which were made of wax, the Blue Amberols were made of a durable celluloid material. The Blue Amberol records had such a hard surface that it was recommended that a special reproducer with a diamond stylus be used rather than the sapphire stylus which was fitted to the model 'C', 'H' or 'K' reproducers. (Conversely, a diamond stylus must not be used on wax cylinder records. The records will suffer irreparable damage.) Conversion kits were made available by Edison for replacing the entire carriage assembly and reproducer with a new one compatible with the new Blue Amberol records. The new reproducer was known as the *DIAMOND 'B'*.

The Diamond 'B' reproducer was designed to play Edison's new Blue Amberol celluloid records.

Although disk records were obviously a more acceptable form than cylinder records, Edison continued to manufacture cylinder phonographs through most of 1929. He also continued to manufacture cylinder records for those customers who previously had purchased a cylinder phonograph and were not ready to switch to the disk format.

For extensive information on the Edison cylinder phonographs, please refer to the George Frow book The Edison Cylinder Phonograph Companion. This book will provide the most detailed and most complete information on the Edison cylinder phonographs.

BERLINER AND THE
VICTOR TALKING MACHINE COMPANY

In the mid-1880's, Emile Berliner saw that one of the main problems with cylinder records was the difficulty in duplicating from a master recording. With this in mind, he developed a flat, round format known as a disk. Where, on a cylinder record, the grooves ran around the circumference of the cylinder from left to right, the disk consisted of a continuous groove beginning at the outside of the platter in a spiral, with the loops growing ever smaller in diameter as they approached the center of the platter.

As referred to previously, the master disks were of zinc, and a stamper (a 'negative' of the master and not playable) made from the master was used to press commercial records. The earliest disk records that Berliner produced were 5" in diameter and of a hard rubber composition. Later Berliner records were increased to 7" in diameter, using the same rubber material. Today the 5" records are extremely rare, and the 7" Berliner's almost as rare and very collectible.

To play these disks, Berliner developed a machine which would track across the flat records. Named the *GRAMOPHONE*, the earliest of these machines were hand operated. Turning the crank would turn the turntable and the record. It was a difficult task to maintain a constant and correct speed, and the sound often ranged from a s-l-o-w to a very fast pace.

The earliest Berliner disk Gramophones did not have a spring motor. The crank used a belt to drive the turntable. It was extremely difficult to regulate speed.

Berliner recognized this problem and sought to develop a gramophone with a motor that would provide a constant speed and ease of use. After several false starts, he contacted a machinist named Eldridge R. Johnson of Camden, NJ. Johnson, using a modified spring motor from a sewing machine, developed the spring disk motor. So popular did these spring-wound gramophones become that Johnson gave up his other interests and concentrated only on the gramophone business. The earliest of these machines were under the Berliner name.

One of the most famous and most recognized was the spring motor *GRAMOPHONE* used in the now famous painting of "His Master's Voice" by Francis Barraud. The puzzled terrier, *Nipper*, seems to be searching for the elusive source of the voice from the gramophone.

The Berliner Gramophone was used in the painting "His Master's Voice". This painting became famous worldwide, and this model Gramophone is fondly called 'The Trademark'.

The later machines produced by this company bore the name of the manufacturer as the 'ELDRIDGE R. JOHNSON COMPANY'. In 1901, after a long and bitter legal battle over patent infringement, the courts found in favor of Johnson over Frank Seaman, a former distributor of Berliner products. So pleased was Johnson that he was found victorious that he renamed his company the 'VICTOR TALKING MACHINE COMPANY' (VTMC), a name that remained synonymous with disk phonographs and talking machines for decades to come. Emile Berliner continued his development and experiments with records and retained his affiliation with the VTMC, which had acquired his disk recording, mass production and reproduction patents as well as the "His Master's Voice" trademark. Members of the Berliner family actively participated in the Victor company until RCA acquired the company in 1929.

The first Berliner machines utilized letters to identify models: *TYPE A* and *TYPE B*, the Eldridge Johnson model was known as a *TYPE C*, and

The Johnson "C" with front–mount horn

subsequent Victor Talking Machines at first carried letter designations (Type P, Type E, Type M, etc.), but were often known by names. The Type E, for example, was the *MONARCH JUNIOR*, the Type R was the *ROYAL*, the MS was the *MONARCH SPECIAL*. These early machines were 'front-mount' talking machines. The horn was supported from the front of the machine, and the soundbox (or reproducer) was attached directly to the small end of the horn.

In 1903 Johnson introduced his 'rear-mount' phonographs, and offered them with the regal sounding names of *"VICTOR the FIRST"*, *"VICTOR the SECOND"* through *"VICTOR the SIXTH"*. Popularly though, these machines were designated on the identification tag as a "Vic I", "Vic II",...

and "Vic VI". Prices for these machines when they were introduced were "Vic 1" at $22, "Vic 2" at $30, "Vic 3" at $40, "Vic 4" at $50, "Vic 5" at $60 and the luxurious "Vic 6" with a large mahogany cabinet, gold plated tone arm and hardware and gold plated finials on the corner columns at a whopping $100! For a complete description and pictures of virtually every Victor and Victrola phonograph manufactured up to 1929, we recommend that you see Robert Baumbach's book *Look for the Dog*.

Victor the Sixth, or popularly called just a Vic 6, was the most luxurious of the regular Victor line.

By about 1906 the phonograph was less a curiosity than a recognized home appliance. Responding to the public's desire for the phonograph to be less conspicuous in the home, Victor introduced a complete piece of furniture housing what today would be known as a home-entertainment center. The phonograph mechanism was built into a mahogany cabinet, and the horn, rather than coming up and out, was designed to go down into the cabinet and to be concealed by two small doors. The lower portion of the cabinet (and the space to the sides of the horn) were devoted to record storage. Johnson called this new item the Victor-Victr*ola*, adding the suffix OLA to designate the concealed horn phonograph. The very first models were known by the rather unwieldy designation of model *VTLA*. The success of the Victrola was substantial, and Johnson, utilizing the same motor, tone arm and sound box of his top-of-the-line "Vic 6", designated this Victrola a *VV-XVI*.

The first concealed-horn phonograph produced by the Victor Company was called the 'VTLA'.

The Victrolas all carried the model designation of VV- followed by their model number. Eventually a complete line of Victrolas was available, beginning with the lidless, tabletop *VV-IV* up through the super-deluxe *VV-XVII* (this machine was available in 1919 with a Japanese decorated lacquer finish and an electric motor at a cost of $615!) and several special order period Victrolas for the very affluent. Beginning in the early 1920's Victor restructured its line and prices and introduced a new series of Victrolas, returning to traditional numbering such as VV-50, VV-100, VV-300, etc. At the same time they introduced a line of 'lowboy' or console cabinets in addition to the traditional uprights. Many of these new Victrolas could be equipped with an electric motor (at an extra charge) and they were designated "VE-" rather than "VV-".

In the mid-1920's, responding to a waning interest in Victrolas and increased technology of 'electrical recording' (utilizing electric devices to increase volume and clarity in the recording), Victor introduced its all new *"ORTHOPHONIC"* Victrola. The technology of the enclosed horn was vastly improved as was the new *Orthophonic* soundbox. A full line of Orthophonic Victrolas was introduced from 1925, but amplification and the playback of the records was still of the acoustical method. The great success of electric amplification spurred Victor to add this feature to their new line of *"Electrolas"* starting in 1926. Many of the new Electrolas also had record changers and/or radios built into the chassis.

In 1929, the Radio Corporation of America, pioneers in the development and use of the vacuum tube and electrical amplification, purchased the Victor Talking Machine Company. The new company officially became RCA-Victor.

There are several excellent books which will provide additional information on Emile Berliner, Eldridge Johnson, the Victor Talking Machine Company and RCA-Victor. Please see the volume His Master's Voice Was Eldridge R. Johnson by E.R. Fenimore Johnson, his son. It is perhaps the most complete biography of Eldridge Johnson. His Master's Voice in America by Fred Barnum is a complete and very fine history of the various companies based in Camden, NJ. Beginning with Berliner, through Johnson and his companies to the acquisition of the VTMC by RCA and finally the purchase of RCA by General Electric, this book is superb. Look for the Dog by Robert Baumbach is the best available chronology of the variety of phonographs manufactured by the Victor Talking Machine Company. All of the models up to 1929 are pictured and discussed in it. Also see Roland Gelatt's 1955 book The Fabulous Phonograph for additional information on the early Berliner/Johnson years. Although out of print, this book may be available from your local library.

THE COLUMBIA GRAPHOPHONE COMPANY

One of the most colorful and interesting stories in the development of the talking machine industry is that of Columbia.

Originally the company was founded by Alexander Graham Bell. Bell, having won the prestigious and lucrative prize, the Volta, for his invention of the telephone, now had the funds to pursue his long-time interest in sound. Bell's cousin, Chichester Bell, and a family friend, Charles Sumner Tainter, were brought over from England in the early 1880's to start the Volta Laboratories. One of their first projects was to obtain one of Edison's tinfoil phonographs and study it.

From their examination of Edison's machine, Bell and Tainter made several improvements, the most important being the substitution of a slender wax cylinder for the hard-to-handle tinfoil. They also incorporated a spring-wound motor into their talking machine. Edison was using the word 'Phonograph' to describe his machine, so Bell and Tainter merely switched syllables around and called their machine a *GRAPHOPHONE*. They formed a new company, The American Graphophone Company, for the manufacture of the machines.

Initially the graphophone (and the phonograph too, for that matter) were not sold, but leased. Since they were intended as business machines, it was companies rather than individuals that leased them. The United States was divided up into marketing territories, and agents for the American Graphophone Company were appointed for each territory. The most successful of these agents were those covering the District of Columbia, home of the federal government. The federal government offices saw potential in automatic transcribing machines and leased many of them for government use. So successful was this group of local leasing agents, that over the years they took over the management and operation of the parent company and renamed it **The Columbia Phonograph Company**.

Columbia (and the entire Bell/Tainter group before it) is noted in retrospect for several very important contributions to the talking machine industry. The wax cylinder is theirs, and it was their presence and research on their Graphophone that caused Edison to again work on his phonograph, making tremendous improvements to it.

Columbia introduced a phonograph for sale designed to satisfy even the tightest household budget: the *MODEL Q*.

The **MODEL Q** was originally introduced to retail at $5! It was a simple machine: key-wound, no cabinet and the entire mechanism was exposed, but the price was right. It was the **MODEL Q** which was the catalyst for Edison to introduce the **GEM** at a retail of $7.50.

The Columbia MODEL Q was an inexpensive $5.00 when first introduced.

Columbia followed its initial success with a slightly better machine, the **MODEL B** which had a double mainspring and a simple oak cabinet. The **MODEL B** retailed at $10. During the late 1890's and early 1900's, a $10 gold piece was known as an 'Eagle' because of the image

The Columbia MODEL B was known as 'The Eagle'. It could be purchased for a $10 gold piece called an Eagle.

minted onto it. Because the **MODEL B** could be purchased for one of those gold pieces, it was nicknamed 'The Eagle', a designation that has stuck. Today that model is more often referred to as an Eagle than as a **MODEL B**.

The Columbia MODEL AB had two mandrels.

Columbia introduced a complete line of cylinder phonographs, from their very inexpensive and very popular **MODEL Q** to rather unusual machines like the **MODEL AB** (the MacDonald) which had a movable mandrel shaft and came with two mandrels: one for the standard size cylinder records, and the other to play the 5" Grand cylinders. Over the few years that Columbia was in the cylinder phonograph business, they had over two dozen models, designated by initials: "Q", "AB", "AT", "AG", "BK", etc. Columbia even introduced a special model, the **MODEL BC**, known as the Twentieth Century with a special 6" long mandrel to play special long-playing as well as regular cylinders. Today these 6" long cylinders are quite rare and collectible.

Columbia was also an aggressive marketing company, and it was the major source for several of the mail-order catalogs, like the young Sears & Roebuck and Montgomery Ward. They also manufactured phonographs for these and other companies under a private label program, and today Columbia-manufactured phonographs with names like *'HARVARD'* and *'LAKESIDE'* turn up with some regularity.

With the almost immediate success of Berliner's disk record and gramophone, Columbia copied their lead and introduced a line of disk phonographs. They realized that the market demand for the disk would outlast the cylinder, and they quickly abandoned the cylinder record business in favor of the disk format.

Following their success with cylinder phonographs, Columbia developed a full line of disk phonographs at virtually all pricepoints. They ranged from very simple early phonographs, with plain oak cabinets and front-mounted horns, to the luxurious, quadruple mainspring *MODEL BY* Improved Imperial with a mahogany cabinet, curved sides and a matching mahogany Music Master horn.

Columbia's luxurious MODEL BY, with a mahogany Music Master horn.

Initially the Columbia disk phonograph was an outside horn type, but when Victor introduced the concealed horn Victrola, Columbia followed with their own concealed horn Grafonola. Columbia was extremely aware of the public's desire for decorative cabinetry, and introduced their Grafonola housed in a variety of unusual cabinets: a miniature baby grand piano cabinet, a desk, a table and they even supplied mechanisms to the Regina Music Box Company for a combination which was an interchangeable music box and disk phonograph.

Columbia's answer to Victor's Victrola:
The GRAFONOLA

Through aggressive marketing, Columbia achieved major importance in the talking machine industry, rapidly becoming one of the 'Big Three' phonograph companies.

One of the most important contributions of Columbia to the record industry was the development of the Double-Sided Disk record. Formerly, disk records were only recorded on one side. Columbia's Double Disk was marketed as "two records at a few cents above the price of one...", "double value for your money, plain as daylight...", and "double disk, double value, double wear, double everything except price"(1). The new Columbia double-disk records retailed as low as 65¢ each, with certain Grand Opera selections available for as much as $7.50 per record! Victor followed the Columbia lead, but Victor retained their one-sided records under their Red-Seal (classical) label into the mid-1920's as a statement of pride and individuality.

Columbia introduced the Double-Sided disk record.

(1) From the Columbia Phonograph Company's Special Demonstration Double Disk record, circa 1909/1910.

EDISON'S DISC PHONOGRAPHS

The Victor Talking Machine Company and the Columbia Graphophone Company had both been successful with the marketing of the disk format records and machines. The public had responded well and the disk was surely overtaking the cylinder in popularity. Although Edison had conceived of a disk format some thirty years earlier, he retained the cylinder as his format. The reasoning that he used was sound: with a cylinder record the speed of the record under the stylus was constant, and so the distortion was minimized. The Vertical or Hill-and-Dale method of recording was his by patent and he felt the sound quality was superior. And though for many years he resisted, the increased pressure from the evident success of Victor and Columbia forced Edison to introduce a disc record and phonograph.

Because he felt so strongly in favor of the hill-and-dale recording, he retained it. The Edison disc records utilized the same recording technology as his cylinders. With hill-and-dale though, warpage of the record would cause distortion. To solve this problem Edison made his records a full 1/4" thick. They would not warp and remained flat on the turntable.

The material that he used for his records ranged from a woodflour and lampblack combination, to a chalk base, to a clay base and all had a 'condensite' varnish finish and a very hard surface. To minimize stylus wear, Edison used a precision ground diamond as the stylus. The point was virtually permanent, suffering primarily from careless handling and not from record wear. Even today, many original diamond styli are in use with no need to replace them. (These special diamond disc records should be cleaned with alcohol and not water.)

Edison believed in technological superiority, and his disc phonograph was no exception. In designing this new machine, he discarded many of the techniques used by his competition (he was also limited by patent restrictions enforced by the competition). The Victor and Columbia phonographs used the spiral groove to move the stylus from the outside to the inside of the spiral with a great deal of record wear resulting. Edison incorporated a worm-gear and comb which was a mechanical method of transporting the stylus and reproducer, that were attached to the tone arm and enclosed horn assembly, across the surface of the record. Because the weight of the tone arm and reproducer were mechanically supported, the stylus was allowed to 'float' over the grooves, imparting minimum wear to the record itself.

Edison's $6,000 disc phonograph. This at a time when salaries were low (see Appendix I, page 63).

Initially Edison used the Amberola cabinet for his disc phonographs, but he soon broadened the selection with pricepoints to fit every budget. His least expensive Diamond Disc Phonograph was the *A-60*, a simple affair costing only $60. The prices soon ranged to $80, $100, $150 and all the way up to a special order period cabinet selling for $6,000! The mechanical principles were the same, and the mechanisms themselves varied very little despite major retail price variations. The major difference in mechanics occurred at the $250 price level. Phonographs retailing at or above $250 used a double mainspring and a large number 250 horn. Each machine meeting these standards was designated an "Official Laboratory Model" and carried a gold-plated medallion attesting to that fact.

The "Official Laboratory Model Medallion".

At $250 one of the Edison 'Laboratory Model' Diamond Disc Phonographs.

The earliest of the Diamond Disc records had an embossed label, black on a black record. The titles were often difficult to read. Later records, known as *Recreations* had a black and white paper label affixed to each side of the record. Because of the special vertical method of recording, the Edison Diamond Disc records could not be satisfactorily played on a traditional disk phonograph such as a Victor or Columbia. Nor could the Victor or Columbia records be played with the Edison Diamond stylus.

Several independent companies manufactured adaptors to fit Edison's disc phonographs which would allow them to play Victor, Columbia and similar laterally recorded records. Edison himself marketed such an adaptor, but soon realized that he was supporting the competition's sale of records.

Special adaptors were available to play 78 RPM records on a Diamond Disc phonograph.

Conversely, jeweled styli were marketed to allow Edison records to be played on other types of phonographs. And several manufacturers, most notably Brunswick, designed a multi-position reproducer which could be adjusted to play traditional lateral-cut records as well as Edison's vertical-cut records.

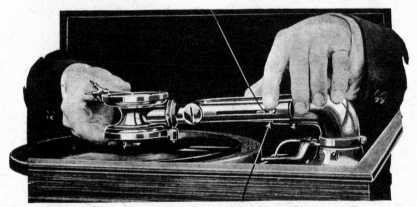

Brunswick reproducer played 78's, Pathes and Edison Diamond Discs records. By moving the reproducer into a variety of positions, three different types of records could be played with the one phonograph.

George Frow has written a companion volume to his cylinder phonograph book entitled The Edison Disc Phonographs and the Diamond Discs. This book details the reasons and motivations for Edison developing his disc record and pictures all of the disc model phonographs, original cost and details of their manufacture.

OTHER BRANDS OF PHONOGRAPHS

When the phonograph was first introduced to the public, people scoffed. They didn't believe that a machine could talk. They kept searching for a very clever ventriloquist. Then they questioned the worth of a phonograph. Who would want one in their home? What would they use it for? And why bother?

Despite the skeptics, the phonograph <u>did</u> become successful. So successful, in fact, that new manufacturers all wanted to jump on the bandwagon and get into the phonograph business. In the early years, the number of manufacturers was very limited, the technology new and the patents extremely restrictive. Nevertheless, several new manufacturers saw a future in phonographs and wanted to be part of the growing field.

The Columbia Phonograph Company owned many patents and was very protective of its business. So when several of these new companies approached Columbia, they answered "Yes, you can manufacture phonographs, but they *cannot play our regular disk records!*" These new companies had to develop ways of getting around Columbia's restrictions. Five notable companies did so: The Standard Talking Machine Co., the United Phonograph Co., the Harmony Phonograph Co., the Aretino Phonograph Co. and Busy Bee (Hawthorne & Sheble). These companies marketed phonographs which would not play regular Columbia records. They did this by modifying the center spindle.

The Standard Talking Machine Model 'A': a simple oak cabinet and red horn.

The *STANDARD* used an oversized spindle of 9/16", the *HARMONY* used a 3/4" spindle, *UNITED* a 1 7/16" spindle, *ARETINO* a giant <u>3</u>" spindle and *BUSY BEE*, used a normal sized spindle, but had an extra plug screwed onto or cast into the turntable so a regular record would not fit.

Five companies marketed phonographs under their own labels. The machines were generally manufactured by Columbia, but the records had to be of a type which would prevent them from being played on a regular Columbia phonograph. The spindle sizes of the phonographs varied, as did the corresponding center holes: (left to right) top row: Standard – 9/16", Harmony – 3/4"; middle row: United – 1 7/16", Aretino – 3" bottom row: Busy Bee – an extra hole which fit a special plug on the turntable.

Columbia was gracious enough (or shrewd enough) to not only manufacture the phonographs for most of these companies using the unusual spindles and slightly varied cabinets with distinctive labels and brightly colored horns (Standard used a red horn, Harmony a blue and Aretino a green, but often the horns did get switched around and an Aretino will show up with a red horn, or a Harmony with a green, etc.), but they also supplied the records for many of these companies. It is not at all unusual today for a 9/16" hole size **Standard Talking Machine Company** record to turn up with the label missing or torn, revealing a **Columbia** label beneath.

These new brands of phonographs would allow a phonograph dealer or a department store to sell a phonograph which seemed like their own store brand, much like Sears, Roebuck and Company did with their *'HARVARD'* or their *'SILVERTONE'* brands, or as Montgomery Ward did with its *'LAKESIDE'* brand. They could discount their phonographs, or even give them away to draw customers into the store and build business for other departments.

In later years, as the enclosed horn phonograph developed, many new companies jumped into the fray. They would build their own cabinets or have them built for them. They would buy motors, tone arms, reproducers and other parts from independent companies, often switching allegiance when they could make a better buy somewhere else. It is not unusual to find a phonograph 'manufactured' in a small Midwestern town. Often these smaller companies used inexpensive cold-cast parts (today known as white or pot metal). They would do the job, but cold-cast was not as durable as machined parts, and the metal cracked, split and shattered. But they were priced inexpensively and competitively during their day. Today many of these secondary and tertiary brands still play well, but many others show the years' wear and deterioration of the 'pot-metal' parts.

Not all of the secondary manufacturers used shoddy parts or materials. Some phonographs were really well constructed, with beautiful cabinets and excellent mechanisms. Many of these companies stood the test of time and today phonographs like the *BRUNSWICK*, the *SONORA* or the *CHENEY* (to name just a few) are well respected and quite collectible.

GIVEN AWAY ABSOLUTELY FREE
Latest Type Hornless Talking Machine
This Most Wonderful Home Entertainer One to Every Family in this Locality
Gives Joy—Pleasure—Anywhere—Everywhere!

 F R E E

At This Store Only to Our Customers

 F R E E

How to Get One of These Machines Free with $25 Cash Trade

These Machines are given away to ADVERTISE A WELL-KNOWN DISC TALKING MACHINE RECORD. The manufacturers are of the opinion that if several hundred of their machines were placed in that many homes in this vicinity, it would create an enormous demand for their Records.

The instruments are now on display in our windows. Call and see them.

YOU DON'T PAY ONE SINGLE PENNY FOR THIS MACHINE.

Commencing this date a free talking machine coupon will be given with every purchase, according to the amount of your sale. For example: If your purchase amounts to $2.50, you will receive coupons to that amount. You save these; when you have a total amounting to $25.00 worth of coupons, BRING THEM IN AND EXCHANGE THEM FOR A TALKING MACHINE ABSOLUTELY FREE.

THE FAIR DEP'T. STORE :: Lakeview, Ill.

Retail stores would use the phonograph as a advertising item. They even gave them away free. With the purchase of any type of merchandise in the store the customer received an equal amount in coupons. When the customer had enough coupons, he could buy the phonograph.

Source of ad unknown

LANGUAGE PHONOGRAPHS AND SCHOOL MODELS

Initially, Edison saw the phonograph as a business and learning tool. He did not want it to be used for idle entertainment or as a toy. But it was not long before people realized that they could play instruments, sing and create their own entertainment with a phonograph. The industry caved-in to the demands of the people, and musical records, stories and jokes were becoming mainstays of the record company's stock-in-trade.

Several very practical ends were achievable with the phonograph over the years. One of the most evident was the ability to learn a foreign language by listening to an actual voice speaking that language.

Three companies moved to the forefront of the learn-a-language

business: the Cortina Language Record System, the International College of Languages of New York City (the Rosenthal system) and the International Correspondence Schools (I.C.S.) of Scranton, Pennsylvania (originally called the International Textbook Company).

Three companies led the learn-a-language business: International College of Languages, International Correspondence Schools and the Cortina Language Record System.

Professor Cortina used both Edison and Columbia machines for his language system and offered record courses in Spanish, French, Italian and German. In the early 'teens he switched from the cylinder record format to disk records. Cylinder boxes from the Cortina system proclaim "Originators of the Phonograph Method of Teaching Languages".

Dr. R.S. Rosenthal used a Columbia *MODEL QQ* as the gramophone of choice in his language courses. The Columbia *MODEL QQ* was modified to run at a slower than normal speed to give longer playing times to the records. The oak lid for this phonograph has a banner decal proclaiming "The Language Phone - International College of Languages". Rather than a horn, Dr. Rosenthal used a single-earpiece listening tube.

Dr. Rosenthal testing a Language-Record

The International Correspondence Schools used the Edison cylinder phonograph for their language course. It is not uncommon to find an Edison *STANDARD* or *GEM* with a small celluloid plate affixed to it identifying it as from the International Correspondence Schools, Scranton, PA. The Edison *STANDARD*, modified for the International Correspondence Schools had a 'repeater' mechanism added. The accessory allowed the listener (student) to depress a lever, which lifted the carriage assembly and moved it back one groove. The advantage of the repeater was that, when listening to a foreign language, a phrase could be repeated for better understanding. The repeater mechanism did not otherwise interfere with the normal operation of the phonograph.

The I.C.S. Learn-a-Language Phono Outfit.

Essentially these courses were all similar: the student would receive a set of language records of his choice, would follow the textbook, and listen to the records. He would then make a recording (as prescribed by the textbook) which he would mail back to the company in special mailing boxes for grading. The Edison *STANDARD* I.C.S. outfit, for example, would include an Edison *STANDARD* 2 minute phonograph, a 14" black and brass horn, a speaking tube and recorder (for recording lessons and tests in the foreign language), blank records with special mailing boxes, a set of language records and a textbook.

In 1904, Sears, Roebuck & Co. offered a disk record foreign language course taught by Graphophone. The $15 course consisted of twelve disk records and an instruction book. It was available in either French or Spanish.   Sears did state, though, that they had in preparation courses in German, Italian and English. The course was produced by the United States School of Language.

Through a special arrangement with the United States School of Languages, we can furnish text books and disc records for complete courses in the study of the French and the Spanish languages. The graphophone makes the teacher's voice available at all times, and constitutes an ideal method for the acquirement of a foreign language. The reproduction is absolutely perfect, clear, distinct, in fact, equal to the original voice. A complete course consists of the instruction book and a set of twelve records. We can furnish at the present time courses in French and Spanish. We have in preparation courses in German, Italian and English. Any disc graphophone or other talking machine can be used. The records fit any kind of a disc machine. For more complete information regarding the study of foreign languages by means of the talking machine, send for special circular.

No. 21B1600 Complete Course in French, 12 records and text book. Price..............**$15.00**
No. 21B1602 Complete Course in Spanish, 12 records and text book. Price **$15.00**

Phonographs were also finding their way into the school rooms. Both the Victor Company and the Edison Company offered special 'School' model phonographs designed for classroom use.

The Edison *SCHOOL* model was identical to the *OPERA* except that the mechanism was enclosed in a black metal cabinet and was on a wheeled metal stand with four drawers for special Blue Amberol records. This was the only model of outside horn phonograph not eliminated by the company in 1913.

The Victor Talking Machine Company introduced a special oak cabinet phonograph which it called the *SCHOOL HOUSE* Victor and was designated as a Model XXV. The upper cabinet contained the motor, tone arm and turntable assembly. It was supported on four legs and had a folding shelf near the bottom. The special oak lid was removable and would hang on the back of the cabinet when the phonograph was in use. A large smooth oak horn slipped into a socket behind the tone arm and would store on the bottom shelf when not in use. The removable lid did have a lock so that the children could not play the machine unless permitted. A later model *SCHOOL HOUSE* had an enclosed horn and two large rear wheels so that it could easily be moved around the classroom.

Victor was very aggressive in using their phonograph in the schools, and they published a series of books and pamphlets designed to be used with the phonograph as a teaching tool. Titles included *Victor Records Suitable for Use in the Teaching of English Literature, What We Hear in Music, Music Appreciation with the Victrola for Children, Music Appreciation for Little Children, Music Manual for Rural Schools with the Victrola* and others.

Victor literature for use with the Victrola in schools.

The phonograph was not just an educational tool for the schools or for learning a foreign language. Both Victor and Columbia put out a series of records and instructional pamphlets on how to learn the popular dances of the day, complete with spoken instructions on where the feet were to be at any particular time. Popular dances included the Fox Trot, the Tango, the One-Step and the Hesitation.

Victor and Columbia each had their dance experts advocating their particular dance instructions. Columbia had G. Hepburn Wilson, M.B., touted as "The World's Greatest Authority on Modern Dance", and Victor had the very popular Vernon and Irene Castle as their spokespersons (or dancepersons).

Physical Health was another important subject for the phonograph record. At least three sets of Exercise records were produced: Victor Records for Health Exercises (three double- sided records and complete illustrations of the exercises), Walter Camp's "Daily Dozen" Health Building Exercises (10 sides plus an instruction manual <u>and</u> a full-color chart of the muscles and organs and the effects of exercise on them) and finally Wallace's Exercises (a five record set complete with photographs of the exercises <u>and</u> a rather forthright booklet, *A Woman's Birthright*, describing specific exercise needs for women and many personal endorsements of Wallace's program).

Wallace's program included a booklet targeting the modern 1920's woman.

While not specifically a learning tool, during the early twentieth century the phonograph became a means of actually hearing the world leaders, statesmen and other notables speak. Columbia was foremost with their **Nation's Forum** label. These records consisted of one side of a disk record which was a recorded speech by such famous names as President Warren G. Harding, Dr. Stephen Wise, Senator Henry Cabot Lodge or Senator Miles Poindexter. The reverse side was generally a military march.

The voice of Pope Pius XI

The **Vatican Record Company**, an unusual label, had a speech by Pope Pius XI in Latin, with the reverse side of the record an English translation.

Victor put out a special-label recording of Queen Mary and King George of England. The label had photographs of both monarchs. Many well known literary and show-business names recorded in the spoken-voice format: poet James Whitcomb Riley, young Jackie Coogan on the Boy Scout label, Commander Robert Peary on the discovery of the North Pole, Benito Mussolini, directed to the Italian-speaking people of North America, Will Rogers and others all recorded so that the world could hear their voices and their messages. Even Thomas Edison recorded a message to our allies after World War I thanking them for their efforts and support.

Victor made it possible to hear the actual voices of the British sovereigns Queen Mary and King George V. This special label included photographs of the monarchs.

PORTABLE PHONOGRAPHS AND NOVELTY PHONOGRAPHS

Close your eyes and picture this: the year is 1924, it is a warm summer evening, a gentle breeze is blowing and the moon is shining brightly. The young man is wearing his best jacket and a brand new straw skimmer hat, and his special girl is in her newest pink gingham dress. The porch light is out, and the swing is squeaking very quietly. What's missing from this perfect picture? Why music of course!

Or it's Sunday afternoon, the sun is bright and the sky is filled with fluffy white clouds. He has paddled the canoe into the middle of the lake where other couples are also 'pitching woo'. What's missing from this picture? Again: music!

Or it's the annual company picnic, and the weather couldn't be more perfect. The young people have paired off and are sitting together. We need MUSIC!

And we've got it! No, it was impractical to carry a floor model Victrola out in the canoe, but several very clever manufacturers solved that problem. They developed very small and very portable phonographs, often small enough to fit right inside the picnic basket next to the egg salad sandwiches. Add a small stack of disk records, and the picture is complete.

Victor and Columbia and several other companies introduced portable phonographs which were substantial in size. Some were 15" wide, 18" deep and 8" high. But the miniature portable was a lot less cumbersome.

Brands like the Swiss-made **MIKIPHONE**, the boxy **CAMERAPHONE** (not bigger than a Kodak), or the **PETER PAN** were all ultra small. The **MIKIPHONE** collapsed into a round case 5" in diameter and only 2" thick. Another model, the **COLIBRI** when closed was only 3 1/2" x 4 1/2" x 3 1/2". But they all would play a ten inch record.

The MIKIPHONE: one of the smallest portable phonographs.

Because of their size and portability the portable phonographs became very popular, and because they completed the picture of the guy and his special girl, they were often fondly called 'Courting phonographs'.

In the home the phonograph was no longer just a curiosity. It had been accepted as part of almost every home. But for those who wanted something really different in a phonograph, the manufacturers obliged.

Columbia had some very creative styling in their Grafonola cabinets. This style, the 'ADAM' was hand-decorated in a Chinese motif.

The cabinetry became very ornate: lacquered and hand-painted finishes could be special ordered; wicker was popular and period styling was a mainstay of any large phonograph company's catalog. Edison, for instance, had a William and Mary, a Chippendale, a Hepplewhite and a Queen Anne cabinet. He even went so far as to introduce a special collection of Art Models, including a large (84" high, 76" wide) French Gothic Art Model which retailed at a whopping $6,000!

Several manufacturers introduced phonographs in cabinets which appeared to be something else entirely. There were occasional tables, there were desks, there were lamps and there were even small baby grand pianos. All had phonographs cleverly built into the cabinets.

The FAIRY LAMP was a combination phonograph and an electric table lamp.

COIN–IN–THE–SLOT PHONOGRAPHS

It is interesting to note that although the original intent of the phonograph was educational, fortunes were made by operators of 'Phonograph Arcades'.

From about 1889 or 1890 very savvy businessmen would obtain specially modified phonographs (and graphophones) which could be operated only when a suitable coin was deposited into a slot. At first these talking machines were leased to operators. They would be placed in saloons, hotel lobbies, or arcades where row upon row of these machines would offer the patron a choice of recorded entertainment.

During the 1890's a talking machine of any type was a bit of a rarity, and people were enthralled with the opportunity to actually hear a voice from a machine, just for putting a Coin-In-The-Slot!

A very clever story-teller on records was Cal Stewart. His "Uncle Josh" character was a rather simple Yankee who spent his days in a series of (mis)adventures. Cal Stewart recorded his famous "Uncle Josh Stories" on both cylinders and disk. A very early recording (on a brown wax cylinder record) finds Uncle Josh in a hotel. Not being familiar with a phonograph, he decides to give it a try. Not quite knowing what to expect, his first experience turns out to be rather frustrating:

"...kind of a little round machine setting down in the office. I think they called it a "Funnygraph" near as I can remember. I got to looking at it and I noticed on it where it said drop a nickel in the place where you put the nickel and put the tubes in your ears and you hear a fellow sing a song or make a speech or do something of that kind. Well I wanted to hear what it had to say, so I dropped a nickel in, put the tubes in my ears and just then there was a band commenced to playing around there someplace and I went out to see where the band was. I don't know what became of it. I couldn't find it anyplace and when I came back in the machine was stopped and I didn't get to hear what it had to say...." [1]

Reports from several of the Coin-In-The-Slot phonograph operators was phenomenal. One company (Louisiana Phonograph Company) had reported that in two months one machine had taken in one thousand dollars (in nickels); another company (Missouri Phonograph Company) had about 50 machines on location, one of which took in about $100 per week!

---

(1) From "Uncle Josh Weathersby's Troubles in a Hotel", recorded by Cal Stewart on Edison Cylinder Records

The very first coin-operated talking machines operated with electricity from a battery. As spring-wound graphophones and phonographs were developed they replaced the battery operated type. These first 'Coin-Ops' were single record, single play machines. They contained only one record each, and to hear a different song meant moving to a different machine.

Columbia had two basic model Coin-In-The-Slot Graphophones: the early *TYPE AS* (using the **Type A** motor) and the slightly later *TYPE BS*, a **TYPE B (EAGLE)** motor in a curved front-glass case.

The turn-of-the-century Coin-Operated Graphophone would play one 2-minute cylinder for just a nickel.

Edison manufactured a broad range of these Coin-In-The-Slot machines, utilizing the Edison *CLASS 'M'* or *CLASS 'E'* phonographs, first with battery power, later with rechargeable (D.C.) batteries, and then line current. Beginning in 1898, spring-driven motor phonographs were used. A variety of Coin-Ops appeared using the **GEM** motor (known as the *BIJOU*), the **HOME** motor (the *'H' Coin-In-The-Slot*), the **STANDARD** (the *EXCELSIOR*)

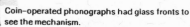

and the **CONCERT** (known as the *CLIMAX*). Each of these machines was fitted into a fancy cabinet with a see-through glass panel. The customer could then watch the mechanism operate as he dropped his coin into the slot.

By about 1900 new Coin-In-The-Slot phonographs were being developed which played multiple selections. The *MULTIPLEX* was a five record attachment for the Edison phonograph; the *MULTIPHONE* held 24 cylinders on a ferris-wheel shaped assembly.

Coin-operated phonographs had glass fronts to see the mechanism.

The **CONCERTOPHONE** held 25 Columbia 6" long 'Twentieth-Century' type cylinders. The Regina Hexaphone offered the user a choice of six different titles on the new Edison Blue Amberol records for the greatest clarity and volume. The Regina Company offered the Hexaphone as "A Musical Money Maker". They advertised these machines as: [2]

> A Musical Entertainer
> An Automatic Cabaret
> An Orchestra for Dances
> Amuses Your Customers
> Draws New Trade
> Pays for Itself

These early inventions led to the later development of coin-operated multi-play disk phonographs. Early leaders in this field were Gabel, Mills, Seeburg and Wurlitzer. Today several of these companies are noted for their superb 'Jukeboxes'.

Today almost all of the early Coin-In-The-Slot phonographs are extremely collectible; the earlier the model, the more desirable today.

Much of the information in this chapter comes from <u>The Edison Cylinder Phonograph 1877-1929</u> by George Frow, and <u>From Tin Foil to Stereo</u> by Read and Welch. See Frow's <u>The Edison Cylinder Phonograph Companion</u> Book for a complete description and details on the Edison Coin-Slot phonographs themselves.

(2) from a Regina Hexaphone sales brochure distributed by one of the distribution subsidiaries, the New England Automatic Amusement Co.

CHILDREN'S PHONOGRAPHS, NOVELTIES & ACCESSORIES

Just take a walk through any toy store today. It is a grown-up world in miniature: lawn mowers, hammers, screwdrivers, kitchen stoves, supermarket shopping carts, all of the things that we as grown-ups use on a regular basis and which our children want to emulate.

Seventy-five years ago it wasn't much different. Children then too, looked at what their parents had and did, and the little ones wanted the same, just kid-size.

The toy shops were chocked full of toys: dolls, wind-up toys, books and crayons, scooters and phonographs. Yes, phonographs! Well why not? Wasn't that the big household purchase of the time? By the late 'teens and 1920's almost every home had a phonograph. And on an evening when company stopped in, Father would bring out the newest records which he picked up that very afternoon, put a fresh needle in and crank up the talking machine....

And if a child was <u>very</u> good, he might even be allowed to help crank the phonograph. What a thrill! So wasn't it natural that the children would want one for themselves? Several companies introduced actual working phonographs, some with small external horns, others with horns concealed within the cabinet. All with little soundboxes and most

Children's phonos were often brightly colored.

with 5" turntables. Most were brightly painted in red or green or blue, and some even had decals of popular nursery rhyme characters.

The records were specially designed for the little folks, with popular nursery rhymes (*"One, Two, Buckle My Shoe..."*, or *"Mary Had a Little Lamb"*). Some were less frivolous (*"Now I Lay Me Down To Sleep..."*) and some were even educational. One of the most popular which taught

The Columbia Bubble Books had read-along stories and records.

as well as entertained was a read along book with several records. These were called "Bubble Books" and there was an entire series of them. Each Bubble Book had a theme: there was "The Funny Froggie Bubble Book", "The Pet Bubble Book", "The Animal Bubble Book" and about nine others.

Some children's phonographs didn't look like phonographs at all. Previously we mentioned Edison's Phonograph Doll, a bisque head and tin body with a primitive phonograph enclosed. It wasn't cuddly and generally didn't work. But it was the first. By the mid-1920's The Mae Starr talking doll and the Madame Hendren talking doll were both very popular. Although both dolls used the identical phonograph mechanism, they were different sized dolls. Each doll came with a selection of six interchangeable cylinder doll records. Access to the phonograph to change the records was in the back, a winding crank extended out of the right hip, and the small horn was enclosed

The Mae Starr Phonograph doll was one of several styles in the 1920's.

in the body with the sound emerging from the doll's belly. These dolls had cloth bodies and so were more cuddly than their predecessors. The 'Mae Starr' doll was even a popular premium prize for kids. For example, those kids selling a certain number of newspaper subscriptions to their friends, family and neighbors might get to select 'The Mae Starr' doll as a prize.

The kids weren't the only ones playing with toys. The kids had the toy phonographs, but the grown-ups had the phonograph toys. If the thrill of hearing a voice coming from the phonograph wasn't enough, folks had to make the experience even more interesting. An entire series of dancing 'dolls' was available to keep rhythm with the phonograph.

The largest and most exotic was 'Siam Soo'. Soo was 12" high, and stood atop the record on a special stand. Part of the stand rested over the edge of the turntable, and part fit over the revolving spindle shaft. As the spindle shaft turned, Soo danced. And boy, did she dance! That little lady knew how to do the Hootchy-Kootchie.

There were smaller dancing dolls: the minstrel dancer, the boxers (they swung punches in time to the music), the dancing couple....

Siam Soo was a dancing doll which operated as the turntable revolved.

And there were other toys for grown ups: a Gramophone Cinema, also known as a kinephone. A revolving disk viewed through stationery holes made the figures on the disk seem to be walking or dancing or running.

IDENTIFYING RECORDS

THE DIFFERENCE BETWEEN
LATERAL RECORDINGS and VERTICAL RECORDINGS

Lateral and vertical recording is essentially a technical subject and will only be addressed in the briefest way possible. Almost all disk records are laterally recorded, with the Edison Diamond Discs the major exception. All cylinder records are vertically recorded.

To best understand the differences, picture, if you will, a phonograph record cut in half, across the grooves, and greatly magnified. The groove will appear as a **U** . Vertical recording, also known as 'Hill and Dale', consists of the groove with the sound impressions on the <u>bottom</u> of the groove. A rounded (not pointed) stylus or needle 'floats' over the tiny hills and valleys of the sound impressions, transmitting them through the reproducer where they are converted to audible sound. Wear to vertical records from the needle is extremely minimal.

A laterally recorded record would have the sound impressions on the <u>side</u> of the groove. The steel needle moves back and forth (laterally) 'feeling' the sound impression and passing them along to the reproducer and then into audible sound.

Purists may argue in favor of one method or the other, but as far as we are concerned, it is enough to realize that a difference does exist, and basically what that difference is.

VERTICAL RECORDING
Heavy line indicates
sound impressions
on the groove bottom

LATERAL RECORDING
Heavy line indicates
sound impressions
on the groove wall

Cylinder Records

Identification of cylinder records is imperative so as not to play the wrong record on the wrong phonograph or with the wrong stylus.

As we discussed earlier, virtually all of the cylinder records are the same physical size, but they are <u>very</u> different in their properties.

All cylinder records are tapered; they will only fit onto the record mandrel in one direction. If the record does not slide on readily, try reversing it. The title end should always face the right hand side of the mandrel. It is not necessary to force a tight fit of the record onto the mandrel. Place the record onto the mandrel so that just minimum friction exists: just enough to keep the record from slipping on the mandrel. Forcing a wax record will almost certainly end in a broken record.

Cylinder records are tapered to fit.

Never handle a cylinder record by the grooves. Hold the record by placing two fingers, the index finger and the middle finger inside the record. Treat the wax records gently. They are fragile and will break if mistreated.

Wax records can be broken by radical changes in temperature. Do not leave them in a closed car for any length of time. Try not to bring wax records from a very warm room into one which is air conditioned. When cleaning wax records put the cleaner on a cloth and <u>not directly onto the record</u>. Let the cleaner come up to room temperature before cleaning. And be gentle! See the end of this section for details on cleaning cylinder records.

Basically there are six types of records which you are most likely to find: brown wax (2 minute), black wax (2 minute), black wax (4 minute), indestructibles (2 or 4 minute) and Blue Amberols (4 minute).

Brown Wax (2 Minute)

The brown wax records are generally the oldest. Only the very earliest had paper title rings on the end; later brown wax cylinders did not have a title on the end: a slip of paper with the title came with the record originally. These are extremely fragile and require extra care in handling and playing. Brown wax records were very susceptible to a mold or fungus which ate at the wax. The records are often covered with a white or gray film or splotches. In most cases the mold is dead, but it has already done the deed: the sound quality will be terrible. If the sound is completely gone, these records can often be shaved, using a special record-shaving device, and used as recording blanks.

Black Wax (2 Minute)

Almost as fragile as the brown wax, black wax records require very careful handling. They were not as liable to be covered with mold, but it did happen. Generally these records cannot be shaved and used as blanks. There were two basic manufacturers of two-minute black wax records: Edison and Columbia. The Edison records can be identified by the name Thomas A. Edison on the title end (the very earliest Gold-Moulded records did not have titles or other identification on the cylinder end). The records are all wax and black on both the outside and inside. The Edison two-minute cylinder records have a beveled edge (cut at an angle) at the title end. The Columbia records are also black inside and out; sometimes they did have a title on the end, often not. They have a flat or non-beveled end.

2 minute wax cylinders often have beveled ends

Indestructible (2 Minute)

Like the Edison and Columbia records, these are black, but the inside is paper or cardboard, often with a metal ring. They are made from a synthetic, not wax, and are quite durable. The record ends are flat and not beveled.

Indestructible Cylinders are not nearly as fragile as wax.

Black Wax (4 Minute)

Generally only made by Edison, these are very easy to identify. They are wax throughout, and the ends are flat, not beveled. The title is white (occasionally yellow or blue titles can be found, often on operatic selections), and somewhere in the title information is the notation **4M** meaning "4 minute". (Two-minute records were not marked two minute; when they were first made there was nothing but two minute so it was not necessary to identify them as such.)

4 minute wax cylinders are marked "4-M".

Indestructible (4 Minute)

Not as common as the 2 minute indestructibles, but they do turn up regularly. These also have a paper or cardboard inner core and a metal

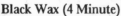

ring. They are marked 4M in the title. Two well-known manufacturers of these records were Columbia and Everlasting.

Blue Amberol (4 Minute)

Manufactured by Edison for use on his Amberola phonographs, or with an 'Amberolized' *STANDARD, HOME* or *TRIUMPH,* the records are almost always blue in color. The blue may range from a light robin's egg blue to a deep, dark blue. A few were colored dark purple. All of the BLUE AMBEROLs have a plaster of Paris core which is most often white. The BLUE AMBEROL records were manufactured of a celluloid and are very durable. But over the years the plaster core often has swelled and it will not fit onto the mandrel. Special Blue Amberol reamers are available to resize these records if they have this problem. CAUTION: when reaming the records, take off just a little at a time. Once it is removed, the plaster cannot be put back again. The ends of these records may be flat or tapered, but they are virtually all 4 minutes. They will <u>not</u> be marked with the 4M.

The plaster core of Blue Amberol cylinders often requires reaming.

Other Cylinder Records

Occasionally an odd size cylinder does turn up. It might be a 'Busy Bee' record (in a pretty yellow box). Although they look just like a regular cylinder, they are a little bigger than standard and will not fit onto anything but a special 'Busy Bee' phonograph.

Edison Concert or Columbia Grand cylinder records are easily recognizable. They are 5" in diameter! These records must be played on a special phonograph with a five inch mandrel. The records are very fragile and quite collectible.

Pathe Salon cylinder records occasionally turn up. They are between the Concert/Grand size and the standard size. They are intended for use only on a Pathe phonograph. (The Pathe Phonograph is a French brand of cylinder phonograph occasionally found in this country.)

Clean cylinder records very carefully. A product similar to *Pledge* furniture polish (in pump bottle, not aerosol) is fine. Spray the cleaner onto a soft, clean rag or towel - check first for buttons or pins in the rag - allow the cleaner to warm up to room temperature, and clean gently around the grooves. Wipe dry and allow to air dry thoroughly. Store in a cylinder record cabinet or in the boxes in which the records came. Replacement boxes are available to store records and keep them clean.

Disk Records

Almost all of the old records that you find will be of the 78 RPM (revolutions per minute) type. We will not concern ourselves with the more modern types of 45 RPM, 33RPM (long play) or 16 RPM. These newer speeds were introduced after 1949 and will not play on a hand-wound phonograph. With very few exceptions, any 78RPM disk record will play on any crank-wound disk phonograph. There is not the confusion surrounding a disk record that there is with a cylinder record. There are a couple of exceptions, and here is where knowing the exceptions is really more important than knowing the rule.

The first exception and the easiest to identify is the Edison Diamond Disc record. Edison records are a full 1/4" thick, 10" in diameter, and will play only on an Edison Diamond Disc phonograph or on a Brunswick Ultona phonograph with the Edison-type stylus. Edison records were vertically recorded. The recording stylus and the playback stylus move in an Up & Down (sometimes called Hill & Dale) motion. The 'sound' is on the bottom of the groove, and the stylus which is rounded and not pointed actually floats over the sound wave impressions on the record. Diamond disk records were recorded to play back at 80 RPM, not 78. (Another exception: in the late 1920's Edison introduced a needle cut or laterally recorded record. These records are quite rare and are clearly marked. They should not be played on a Diamond Disc phonograph.) Edison also introduced a long-playing record in the late '20's. Although vertically recorded, the 28 minute and the 40 minute long play Edisons must be played on a special long-play Edison phonograph with a special long-play reproducer. These records, too, are quite rare.

Several other manufacturers recorded disk records with the vertical recording system. Par-O-Ket and Pathe are two. These records (again, not at all common) should be played with special reproducers which adjust to a different angle of tracking and use a ball-shaped sapphire stylus.

Now back to the normal, everyday 78 RPM disk records. There are literally hundreds of labels of 78's (see Barr's book *The Almost Complete Guide to 78 RPM Records* for an list of labels and recording dates). Some of the more common labels include Victor, Columbia, Decca, Bluebird, Brunswick, Vocalian, etc. These records are <u>laterally</u> recorded. The sound is on the side of the groove rather than the bottom. As the needle passes over the sound impressions, it moves back & forth rather than up & down. These records are designed to be played with a new steel needle (fiber, cactus or thorn will also work), and the needles have to be changed at least every other play. It is the taper of the steel needle and not the point that experiences the wear.

The very earliest disk records were by Berliner and were 5" in diameter. Today these records are extremely rare. Later Berliner introduced a 7" record. When Eldridge Johnson joined Berliner the records became known as Johnson records, and then Victor records. Columbia too, made 7" records (the earliest commercially available Gramophones had only 7" turntables). Later 8" and 9" records were manufactured, but the industry settled on 10" as a standard. Today 95% of the records that you find will be 10" 78 RPM's. 12" 78 RPM records gave additional playing time, and were particularly popular for classical music which required more uninterrupted playing time. Pathe used a 14" record and that, too, is quite rare.

Before you start getting excited because you have a 5" disk record in your collection, many 5" records were made in later years, and they do not have the rarity of a 5" Berliner. Little Wonder records, Bubble Book children's records and many other brands of children's records were 5". In England, during the 'teens, 5", 6" and 7" records were very popular, and they show up quite regularly.

The earlier 78 rpm records were made of a harder material than those made in the late 1930's, '40's and '50's. Often when playing one of the newer records on an early phonograph the needle will tend to drag across the record and even slow to a stop. The newer records were designed to be played with an electrical pick-up which was much lighter than the early acoustical reproducers. If this problem occurs, try playing a record from the 'teens or early '20's. The harder surface allows more weight in needle tracking.

Clean 78 RPM Disc Records by using a mild household detergent. Wet the record (do not soak it) under luke warm water. Work up a good lather on a very soft piece of tee-shirt material or toweling, and rinse thoroughly under clear, fresh water. Towel dry and allow to stand in a dish-rack and air-dry. Do not dry in front of an airconditioner, heater, in direct sun or in the dishwasher.

REPRODUCERS, RECORDERS & NEEDLES

The reproducer is the portion of the phonograph system which converts the signals encoded on the record into audible sound. Virtually all reproducers contain some common parts.

The *stylus* or needle actually moves along the surface of the record, and gently 'feels' the undulations which represent the sound. These undulations cause the stylus to move (up and down on Edison records [vertical recording] or sideways [lateral recording] on Victor, Columbia or similar types of records). This movement is then transferred through a *linkage* system - either a fiber or wire linkage or a needle bar - to the *diaphragm*.

The diaphragm, of mica, metal or specially treated paper, converts the motion from the record into audible sound waves. The sound waves are then channeled through an ever-expanding series of tubes (most often a tone arm into a horn, or directly into the horn) where the volume is increased to listening levels.

In most cases, poor sound is due to either the quality of the record or a problem with the reproducer. A worn or broken stylus or needle, bad gaskets and/or diaphragm or a linkage problem will impair the sound quality.

Numerous 78 Rpm labels exist.

Most of the non-Edison disk records found today are laterally recorded. Examples of these records are Victor, Columbia, Decca, Capitol, Vocalion, Bluebird, etc. These records should be played with a new steel needle. The steel needle is actually softer than the record, and experiences microscopic wear each time it is used. Failure to change the needle at least every other play will result in poor sound and rapid record wear. Needles were (and still are) manufactured in various volumes: Full-tone needles were the loudest, followed by Medium-tone, Soft-tone and finally fiber needles. Edison disc records, and Edison cylinder records (as well as all other types of cylinder records) must be played with a jewel stylus, most often sapphire or diamond. Although these styli do wear, it is not a rapid wear, and these styli can last for generations.

Since it is the taper of the steel needle and not the point that plays the record, wear cannot be seen without a microscope. But the steel needles <u>do</u> wear and must be changed regularly. The jeweled styli are precision ground to fit the groove of the record. They are designed to 'float' over the sound impressions in the bottom of the groove. The jeweled styli, most often sapphires, are semi-permanent and require replacement only after a great deal of use.

The Edison Diamond Disc records and the Edison Blue Amberol records are made of extremely hard materials, and the stylus is a diamond, ground to fit the groove. The black cylinders and brown cylinders are generally of a wax material, and a sapphire stylus is adequate. The hard-surfaced 4-minute Blue Amberol cylinders should be played with a diamond-stylus reproducer. The diamond stylus <u>must not</u> be used on wax cylinders.

Because the two-minute records and the four-minute cylinders are the same physical size, in order to increase the playing time the grooves in the four-minute cylinders must be much narrower than a two-minute. Reproducers for cylinder records are designed to play a specific type of record.

The Edison **MODEL 'C' REPRODUCER** is designed to play <u>ONLY</u> two-minute cylinder records; the Edison **MODEL 'H' REPRODUCER** was designed to play four-minute cylinder records, but it can be used for two-minute play without harming the record (sound will not be up to standard, but it will not harm the record). The Edison **MODEL 'K' REPRODUCER** is designed with both a two-minute and a four-minute stylus. By simply rotating the pointer on the bottom of the reproducer, one can select either the two or four-minute stylus.

"K" reproducers have two styli.

Other types of Edison reproducers are often found: the **MODEL 'B' REPRODUCER** (2 minute only), the **MODEL 'N' REPRODUCER** (4 minute), the **MODEL 'O' REPRODUCER** (2&4 minute combination), the **'R'**. the **'S'** and several other specialty type reproducers. Refer to George Frow's *Edison Cylinder Phonograph Companion* book for detailed explanations.

For the most part, reproducers found on Columbia cylinder phonographs are of three types: a **SMALL FLOATING REPRODUCER**, a **LARGE FLOATING REPRODUCER** and the **LYRIC TYPE REPRODUCER** (named for the distinctive musical lyre shape of the carriage). All three types are generally two-minute only.

The Columbia disk phonographs utilized a variety of reproducers. Most of the outside horn Columbias (and a variety of phonographs manufactured by Columbia for other companies) utilized the *ANALYZING REPRODUCER*. This 2" diameter reproducer was made in a variety of styles. On the Columbia phonographs with rear mounted horns the reproducer screwed onto the tone arm with three screws; on the front-mounted horns, the reproducer had a long 'throat' which fit into an elbow on the end of the horn itself. Some of the ANALYZING reproducers used a thumbscrew to secure the needle, and others used a lever-activated spring clamp. Other types of reproducers used on Columbia phonographs included one type that was a 'bayonet mount'. An internal spring held a retaining bullet against a groove on the throat of the reproducer. Removal is achieved by rotating the reproducer approximately 3/4 of a turn and sliding it off. **A WORD OF CAUTION!** Columbia used quite a bit of cast metal or 'pot' metal on their phonographs. Today this metal is extremely fragile. Attempting to force any frozen joint will almost surely result in broken parts.

Edison Disc phonographs often are found with two reproducers: one is the standard Edison *DIAMOND DISC REPRODUCER* and the other an after-market adaptor to allow play of Victor or Columbia type records. The adaptors were made in a variety of styles, but basically all fit into the tone arm of the Edison Disc Phonograph and allow steel needles to be used on Victor or Columbia records. In later years

78 Rpm records cannot be played with a Diamond Disc reproducer.

Edison also manufactured a 'dance' reproducer, the 'EDISONIC' reproducer and a 'long-play reproducer, which would only work with special and quite scarce Edison Long-Play records and an Edison Long-Play phonograph with special internal gearing. Edison reproducers may be removed by turning the knurled ring to the right approximately 3/8" and then sliding the reproducer forward and out.

DO NOT ATTEMPT TO PLAY EDISON DIAMOND DISC RECORDS WITH A STEEL NEEDLE.

Victor reproducers generally present the least confusion. There are three basic types and all are interchangeable. The earliest common Victor reproducer is the *EXHIBITION*. This reproducer was found on both the front-mounted horns and on the rear-mounted horns. The next type is

the *VICTROLA NUMBER 2*, generally found on the later Victrolas. Finally is the *ORTHOPHONIC* reproducer which was specially designed to impart louder, clearer sound to the electrically recorded records which were being marketed in the late 1920's. Victor reproducers on rear-mounted outside horn Victors and all Victrolas are mounted with a key-way lock. To remove the reproducer, hold the 'U' shaped tube in the left hand, and rotate the reproducer  counter-clockwise approximately 1/2" and slide the reproducer off to the right. The *EXHIBITION* reproducer uses a rubber mounting flange on the back side of the reproducer. Often these are broken, cracked or warped and often break when the reproducer is removed from the tone arm. They are easily replaced and should be replaced if broken, warped or hardened.

Other Victor reproducers which are often found on Gramophones include the "Concert", the "Victrola Number 4" and a variety of reproducers made in England as "His Master's Voice" brand, made to fit British 'Victors'.

Perhaps hundreds of other types of reproducers were manufactured during the early years of the phonograph, many of these for specific 'house brands'. Often one these reproducers is found on a brand name phonograph like a Victrola or a Grafonola, and if it does play well, there is no need to replace it.

The final common type of reproducer is the Brunswick. The *ULTONA* was interesting in that it combined several types of record playing abilities into one reproducer.

The first type has only one diaphragm, but will play both lateral cut records (Victor, etc.) and vertical cut records (Pathe) by changing the stylus. The Pathe uses a special ball-shaped sapphire needle.

The second type of Brunswick has two diaphragms. One side is as described above, but the other side is specifically designed to play the Edison Diamond Disc records with a built-in diamond stylus. Both types utilize a series of twists, turns and balances to set the correct stylus in place. It is recommend that you refer to a Brunswick owner's manual for the correct settings.

Recorders

Recorders were devices fitted to a cylinder phonograph so that one might make recordings at home or in the office. Both Edison and Columbia manufactured them for their respective brands.

Rather than a rounded stylus which floated over prerecorded sound, the recorder had a concave stylus which cut grooves in the wax record, moving up & down as the cylinder revolved under it. Speaking (or singing) into a horn would cause the diaphragm, to which the stylus was attached, to vibrate vertically, literally gouging the sound onto the record. After recording, a traditional reproducer would be used to play back the newly recorded record.

Over the years the brown wax records have gotten hard and generally home recording today is less than satisfactory. But owning an original recorder with your phonograph is a nice accessory.

PHONOGRAPH HORNS & HORN SUPPORTS

Strictly speaking, the horn is nothing more than an acoustical amplifier. It is a confined chamber which allows the sound from the reproducer to expand.

Cylinder Phonograph Horns

The earliest cylinder phonographs did not come equipped with horns. Instead they utilized listening tubes, very much like a doctor's stethoscope. Often a number of these tubes would be connected to a single phonograph allowing several people to listen at once, but each listener had to have a set of tubes in order to hear the music. It was not possible to sit comfortably in a room and enjoy the records.

The Edison Company and Columbia each had their own favorite size and shaped horns, and in most cases the horns were interchangeable. For all intents, if the horn fits the machine it will amplify the sound.

On their early cylinder phonographs, Edison preferred the 14" length horn, first in all brass, and later in japanned tin with a brass bell.

Columbia favored both a 10" cone-shaped horn and a 10" belled horn which was completely nickel plated. Today these are the three most common sizes/shapes to turn up with phonographs, and these horns are readily available in excellent reproduction models.

2 Columbia horns.

The Edison phonographs showed a much greater variety of original horns (but Columbia's interest in cylinder phonographs was short-lived, preferring to switch to the disk format). Each model of the Edison phonographs eventually came with a horn designed for that particular model. The Edison **STANDARD** came with a 30", 10 panel black morning glory, the **HOME** with a 32", 11 panel, the **TRIUMPH** with a 33", 12 panel. The **FIRESIDE** with a two-piece 19" horn, and the later **GEM** with a one-piece 19" horn (early **GEM**s used a 10" cone-shaped horn).

During late 1909, the two-piece Cygnet (pronounced "cig'-nit"; French for swan) horn was introduced. Not only was this horn favored for its shape since it was vertical above the phonograph rather than impinging on the space in the room, but the sound was fuller and richer. The metal cygnet bell was available in a 10 panel for the smaller phonographs and in an 11 panel for the larger machines.

The Music Master Horn Company also made wooden horns for the Edison phonographs to fit the cygnet elbow. Although smooth oak was the most common of the wooden horns, exotic woods and decorative horns were also available.

Cygnet is French for swan. Cygnet horns are curved like a swan's neck.

The Edison *OPERA* utilized a horn exclusive to itself. Whereas all of the other horns above 14" in length required a support (either a horn crane or floor stand), the *OPERA*'s cygnet type horn was self-supporting. It used a special elbow (metal, but grained to simulate wood to match the bell). The *OPERA* horn-bell did not have an appliance for attaching it to a crane. Both Oak and Mahogany were used.

The increased weight and balance of phonograph horns over 14" in length required a special support. Several variations of these horn supports were used, and many people over the years, not having the original support readily available, designed their own horn supports.

The Edison horn supports, or cranes were designed for the type of phonograph and size of horn with which they were going to be used. The Edison **STANDARD** horn, the **HOME** horn and the **TRIUMPH** horn were each available with a large, black morning glory shaped horn from the Edison factory. For these phonographs an external horn-crane was supplied. A small bracket was screwed to the underside of the cabinet, out of sight, and into this bracket would fit a support shoe. A molded-in socket would support the main vertical rod of the crane. A second support bracket would fit over the vertical rod, and would fit <u>under</u> the hinged wooden motor frame. A lip on the

A morning glory horn crane.

upper support would be held firmly inside of the wooden cabinet (or in later model phonographs, into a slot in the front top of the cabinet specially designed for this support bracket). The vertical rod had an angle bent into it, terminating with a hollow in the rod and a set screw. A slightly thinner rod would fit into this hollow, and could telescope in or out (depending on the length of the horn used) to give proper support. The smaller rod had a short length of chain at the far end, with an 'S' hook which would attach to a ring at the top of the horn. The small end of the horn would connect to the reproducer with a short length of flexible rubber tubing. When properly connected and balanced, the chain would hang perfectly vertical (straight up and down, not at an angle) and the rubber tube at the small end of the horn would just fit over the reproducer while applying virtually no pressure on the reproducer.

The later model *GEM* and the *FIRESIDE* phonographs had crane sockets cast into the upper works. A smaller crane would fit into this socket to provide support for the 19" horn.

The cygnet horn was supported by a crane that attached to the back of the cabinet with nuts and bolts. A special mount would be bolted to the rear of the cabinet, and a long vertical rod (with a special bend at the upper end) fit into it. A second curved rod fit into the vertical completing a support which simulated the shape of the cygnet horn. A special hanger assembly attached to a special ball on the top of the cygnet horn. This hanger was also used to adjust the horn slightly up or down to give the best fit over the reproducer without putting too much pressure on the reproducer.

Columbia phonographs and 'after-market' horns (that is, horns purchased from a phonograph dealer, and not necessarily the horn which came from the phonograph factory) utilized a variety of cranes. One common type was a clamp bracket which fit under the cabinet, front to back and was tightened with an eye-bolt. Over-tightening often gouged the cabinet front and the wooden bottom frame of the cabinet. Another type also used a clamp assembly which clamped to the front of the cabinet, often causing damage to the cabinet front. Finally, another type of crane fit into the cygnet (or a similar type of back-mounted bracket) and the crane extended over the phonograph to support the horn.

A separate free-standing floor stand was often used as a horn support. This stand generally had folding legs (similar to a folding music stand of today) and a top section curved to fit around the horn. A chain supported the horn. The main problem with the floor stand was that it took up a lot of floor space, and was susceptible to people constantly banging into it. But for the extra-long horns (some as long as 50"), a floor stand was the only means of support.

Regardless of what type of horn support was used, balance was critical. The chain had to be straight up and down and very little pressure was to be placed on the reproducer by the weight of the horn.

After 1913, Edison produced no phonographs with outside horns except for a special school model.

Many after-market or dealer optional horns were available. One could purchase a phonograph and select a solid color or decoratively hand-painted horn to match or coordinate with the home decor. Many of these horns were brightly painted, with the color flowing and blending from the throat of the horn out to the edges which were decorated with flowers. Because they looked like flowers themselves, they were nicknamed 'morning-glory' horns. Most of the after-market horns (but not all) featured convex petal-tips, whereas the Edison horns had concave petal-tips. Also available were larger variations of the tapered horn with the bell. These

Floral horns were painted to highlight the home decor.

'Witches Hat' horn.

have become known as witch's hat horns, because if they are stood on the bell, they resemble that garment. The witch's hat shape was favored by Columbia for their larger horns, and they are found today all the way up to 50" + in length!

A variety of materials were used for the horns. The tins horns had a brighter, brasher sound, the brass and paper-mache softer and mellower and the wood a rich, full sound.

Disk Phonograph Horns

The earliest horns used on the disk phonographs were of the funnel-shaped body with a bell end. Pre-1902 or so, the horns were mounted from the front of the phonograph, and the reproducer fit directly onto the horn itself, or used a metal or leather elbow. Columbia introduced a petal-shaped horn for their front-mount machines, as well as the funnel shape.

With the introduction of the back-mounted horn, the size and weight of the horn became less critical for proper performance. The weight of the horn, the balance and the pressure of the stylus on the record was minimized, resulting in longer-lasting records. Larger tapered horns as well as petal shaped horns began showing up.

The Victor Talking Machine Company designed horns for virtually each of its models, both tapered and petaled, to maximize sound quality and balance the appearance of the phonograph. With very few exceptions, the Victor metal horns were either black with a brass bell, or all black

petaled horns. The introduction of wooden horns (at an additional cost to the customer) allowed for the excellent sound quality as well as the decorativeness of either a smooth or laminated wooden horn. Most of the Victor horns utilized a bayonet-type elbow connector between the horn and the back-mount.

Victor bayonet-type horn mount

Columbia screw-in horn mount

Columbia excelled in their range of horns. Their back-mounted horns were petal shaped, and available in various sizes. Columbia offered not only black horns, but also red, green, blue and nickel-plated! On their larger, ornate cabinets Columbia used beautifully laminated horns of oak and mahogany. Columbia used a screw-base elbow connector between the horn and back-mount.

By the early 1900's the housewife was tiring of the large horns taking up so much of the space in the room and constantly being bumped into by one member of the family or another. Besides that, the horns were considered unsightly, and were dust collectors. The Victor Talking Machine Company heard these complaints and proposed a solution. Rather than have the horn exposed on top or in front of the phonograph, Victor built the entire mechanism, motor, turntable, tone arm and reproducer <u>and</u> horn, into a free standing piece of furniture. Before long, this radical, new, fully enclosed Victor became known as a *VICTROLA.* Columbia soon followed and their *GRAFONOLA* became nearly as popular.

WHAT DID IT COST?

(This section was originally written for George Frow's new edition of his superb book The Edison Cylinder Phonograph Companion in order to give a comparison of the buying power today versus the buying power of almost a century ago.

We are including it as a chapter in this book with the kind permission of Mr. Frow.)

What *Did* It Cost?

In 1900 an Edison *GEM* phonograph cost $7.50, an Edison *STANDARD* $20 and an Edison *CONCERT* $100.!

Today these seem like great bargains, and we all have a spare $7.50 which we would readily spend on a *GEM*, but in 1900....

We cannot properly judge the cost of merchandise from our vantage point so close to the end of the twentieth century. A pound of sugar for 4¢, a dozen eggs for 14¢ and a pound of butter at 24¢ all sound like marvelous bargains.

Further bargains by today's standards include a 2 quart rice double boiler for 25¢, a 4 quart gray enameled coffee pot for 29¢ and the Sunday New York Times for 3¢. Gentlemen's Madras pajamas were available at $1 a suit, ladies summer dresses on sale for $25 or ladies black kidskin boots at $1.00/pair. Gentlemen's suits were to be had at a cost of $20 to $25, with selected styles and fabrics as low as $12. And the well-dressed lady or gentleman could attend a performance of Proctor's Vaudeville Theatre with a choice of tickets at 15¢, 25¢ or 35¢ or take a nine day cruise from New York to Bermuda, all expenses paid for $37.50.

A special sale of building lots in Brooklyn, NY, in the fashionable Flatbush area, just one mile from Prospect Park, ranged between $190 and $590 each, and could be purchased for $2. down, and between $1.50 and $2. weekly at 4% interest. [1] Or a high-stoop house on 183rd Street in New York, just off Broadway, rented for $45/month.

But all of these bargains had to be balanced against the amount of income that a family had. In very few cases did a wife work: once she was married, a young woman left the job market to care for her husband, children and home. The exception might be a husband and wife domestic team, but most did not have families to raise.

For the husband, the workday started at seven or eight o'clock a.m., and continued until 6 o'clock p.m. - Saturday was often only half a day's work! An article in the June 23, 1900 issue of the New York Times dealt with a gentleman who threw his father out of the house, contending that on his $60 a month salary as a painter, he could not afford to take care of his own family and his father too! N.Y. City gardeners were protesting the fact that most earned only $65 a month, while a few selected men in their ranks commanded $75 a month. [2]

Starting salaries for a young man in an office were not nearly so generous. A salary of $30. per month for the first six months was offered and the position probably was filled very quickly. The average salary for a worker at the turn of the century ranged between $40. and $50. per month; that averages between $1.75 - $2.20 per day.

Young ladies had few options open to them in the job market; a (home) nurse commanded $20. a month, a dressmaker $1.50 a day, and a cook $25. a month.

One ad read: "wanted, a refined Protestant seamstress and maid for three young girls; one capable of giving some assistance in management of the household preferred. Wages $20.00...." [3]

Funds for entertainment had to come out of the small monthly earnings. (At $2.20 per day, the least expensive Edison *GEM* at $7.50 cost three-and-a-half days' pay.) And when the cost of an Edison *STANDARD* phonograph equaled approximately three weeks' wages, it became a major purchase (By today's economy, a worker making $40,000/year would have to buy a $1,500 stereo to equal two weeks' wages). Despite that, Edison *GEMS*, Edison *STANDARDS* and *HOMES* did sell, and sell by the thousands! And so did the cylinder records. Each week when *new* records were received by the dealers, owners of Edison phonographs lined up to hear the latest songs or vaudeville routines and buy them for the family's enjoyment at 35¢ each.

1) New York Times June 9th, 1900
2) New York Times October 1st, 1900
3) New York Times June 6th, 1900

NIPPER – THE DOG & GRAMOPHONE

Nipper is the world's most recognized dog, and one of the most recognizable trademarks the world has ever known.

Nipper was a real dog, owned first by Mark Barraud, and later by his brother, Francis. Francis Barraud, an English artist, painted the dog listening to a <u>cylinder</u> phonograph. The original painting done in the late 1890's displeased Barraud. The horn was black and seemed to darken the entire painting. In an effort to brighten the painting, Barraud went to the Gramophone Company in London and requested the use of a brass horn to include in the painting. The Gramophone Company agreed to purchase Barraud's painting if he agreed to paint out the cylinder phonograph and paint in a Gramophone with the new brass horn. In October, 1899, the Gramophone Company (London) purchased the painting and the sole right of "reproducing the picture on trade circulars, catalogues and heading of note paper". (1)

Barraud agreed, and the painting went on to become one of the world's greatest trademarks, making its first appearance in England in January 1900 on the British Record Supplement (2) and in the United States later that same year. Emile Berliner brought a copy of the painting from England after a trip there in May of 1900, and received the US "Trademark for Gramophones" in July. (3)

Eldridge Johnson first used the trademark with Berliner's permission in December, 1900 on the horn of his $3. (Toy) Gramophone, and the trademark first appeared on company letterhead in January, 1901. (4)

Nipper & the Gramophone were first used on company letterhead in 1901.
   Courtesy General Electric Company

But *Nipper* was more than just a stagnant trademark. The charm of the dog captured the public's fancy, and replicas of the dog (often without the Gramophone) were used as display pieces, for advertising and even as giveaways to the public. Today, miniature (3" - 4") chalk Nippers turn up with dealer's names and addresses imprinted on the base.

During the 1930's, '40's and '50's it was common to see a giant (36") Nipper dog in the window of the local 'Victor' store, and smaller dogs perched appealingly on the phonographs within the store. Most of these dogs were manufactured of paper mache by the Old King Cole Company of Ohio. The Old King Cole Company is no more, but its heirs (a plastics technology company) still manufacture *Nipper* of polyethylene for display use by dealers.

Nipper dogs came in a variety of sizes

Large hard rubber Nipper dogs were used on the street in front of a record store, and paper cut-outs of *Nipper* would quickly identify a product as 'Victor'.

In the United States, England and Japan, *Nipper* was a readily identifiable image of the home entertainment business. When the public saw the famous dog, they immediately associated it with phonographs (and later radios, hi-fi, and other home entertainment products.) Even today, in Asia (through the Japan Victor Corporation [JVC] and in England and Europe (through EMI) Nipper remains an active and highly recognizable trademark. In the United States, the trademark use has been curtailed, but is still seen on records, cassette tapes and compact discs marketed through the BMG corporation.

In the early-1940's RCA-Victor introduced a line of children's records, and along with them a cute puppy called *Little Nipper*. *Little Nipper's* image was used generously in decorating the record albums, and the puppy could be found in a number of cute positions on the album cover and on the inner pages.

In the 1980's the RCA-Victor Company was purchased by General Electric, and they acquired the use of the famous trademark. The record division of the company was sold to the German BMG (who acquired

Nipper for use in the Americas only - the European rights are owned by EMI, the descendent company of the British Gramophone Company). The home entertainment division was sold to Thomson Consumer Electronics, a French company. They chose to retain *Nipper*, but eliminated the Gramophone. Along with *Nipper* they again added a perky puppy named *Chipper*. The two dogs are being used in advertisements of the home entertainment products.

Chipper & Nipper:
a new generation

Today *Nipper* has become as collectible as the antique talking machines themselves. Collectors search antique shops, junk shops, flea markets, swap meets and garage sales for anything 'Nipper'. An extremely broad range of 'Nipper' products was produced and marketed either directly by the RCA Victor company or under license from RCA Victor. Over the years many, many items have been produced such as images of the dog, in plaster, ceramic, rubber and fabric, watches, drinking glasses, coffee cups, jewelry, store display pieces, signs, mirrors, banners and flags, letterhead, and matchbooks. Dedicated collectors are industriously searching out these items and collecting them for their own private collections. Even today, a variety of 'Nipper' items are being produced, often in limited quantites for use by the companies. As these new items are produced, many of them are in as much demand as even the older collectibles. And the young puppy, *Chipper* appears to be following in the older dog's footsteps of collectibility.

For detailed information on collecting Nipper, refer to the book The Collectors Guide to 'His Master's Voice' Nipper Souvenirs by Ruth Edge and Leonard Petts. The new 1997 edition lists about 2,000 Nipper items.

(1) The Story of 'Nipper' and the 'His Master's Voice' Picture by Leonard Petts, 1983
(2) Ibid.
(3) "His Master's Voice" in America by Fred Barnum III, 1991
(4) Ibid.

MOST COMMONLY ASKED QUESTIONS ABOUT HAND-CRANKED PHONOGRAPHS

I just got a wind-up phonograph but it doesn't work. Can it be fixed?
Generally, yes! Most phonographs were mass-produced and parts from the same models are interchangeable. Occasionally a maverick brand shows up where parts are not readily available. Often parts can be manufactured for these machines.

I have an old wind-up phonograph. Is it worth anything?
Definitely 'Yes!' How much, we can't say without seeing the machine, but these old phonographs are in dwindling supply, and their value and the demand for them is increasing.

Are vintage, hand-cranked phonographs a good investment?
Like any other investment one of three things can occur: the value can go up, the value can go down, or the value can remain pretty much the same. The value of a collectible is based primarily on the old laws of supply and demand. If the supply is limited and the demand great, your collectible will bring a higher price. But if no one is buying then the value is not going to reflect great gains. As a rule of thumb, the better the original condition of the phonograph, the better the chances for increased value in the future. Also the rarity of the phonograph, the accessories with it and the history of your particular machine can all make the machine more valuable to a prospective buyer, and bring a potentially higher price. Before putting your savings into any type of collectible for investment, we suggest that you discuss your financial situation with your accountant or tax advisor.

I have a wind-up phonograph which has been in the family for years. What should be done to preserve it?
First of all, a thorough cleaning, lubrication and minor adjustments of the motor, tonearm and reproducer are suggested. This will put the mechanics back to first-class order. After that you can decide whether cosmetic restoration is needed or warranted.

I have an Edison cylinder phonograph and the only identification that I can find on it is 'Model C' (or sometimes 'Model H Four-Minute'). What kind is it?
Chances are the Model C or the Model H Four-Minute is the model of the <u>reproducer</u> and not the phonograph. The Model C is a two-minute only reproducer, and the Model H is a four minute reproducer. These reproducers could be used on a variety of different model phonographs. Refer to the chapters on Edison cylinder phonographs or on reproducers for more details. (The same advice goes for a Victrola or Victor model

'Exhibition' or a 'Victrola Number 2'. These are reproducer designations and do not identify the phonograph model.)

How can I tell if my Edison cylinder phonograph plays only two-minute records or both two and four-minute records? Both types fit on the phono.
Physically, both two minute and four minute records are the same size. The four minute capability is determined by whether the reproducer is a four minute type, and whether the internal gearing of the phonograph has a four minute setting. With a four minute record, the stylus moves across the surface of the record half as fast as it does with a two-minute, but both types of records rotate at the same speed. See Chapter 3 "Edison's Cylinder Phonographs" for more details.

How can I tell if my cylinder records are right for my phonograph?
Since both two and four minute cylinder records are the same size, identification can only be made by examining the type of record. Basically there are three types of records: 2-minute wax, 4-minute wax and 4-minute celluloid Blue Amberol. The Blue Amberol is the easiest to identify; the color is almost always blue (with an occasional dark purple). The Edison 2-minute black wax have a chamfered (cut at an angle) title end. The Edison 4-minute black wax have a flat title end and are marked "4-M". See Chapter 12 "Identifying Records" for much more detail.

I have a really old phonograph and the mainspring is broken. Can I get another one?
Yes. mainsprings for virtually all brands of phonographs are available. They are relatively inexpensive. But removing the old mainspring and replacing it is dangerous and very dirty work. It is best left to a professional.

The mainspring on my old wind-up phonograph is getting weak. Can it be replaced?
Generally the problem is not a weak mainspring, but lack of proper lubrication, especially on phonographs with two or more mainsprings. Over the years the oils and grease used to lubricate these machines has dried out, and now the dirt and dried oils act like a glue to slow everything down. Merely oiling the machine will not do the job; it must be professionally disassembled, cleaned and relubricated with the proper oils and greases. Most household oils and automotive greases are **wrong** for an old phonograph and may cause additional problems.

A WORD OF CAUTION: Removing and replacing a mainspring is not a job for the amateur. It can be very dangerous and at the least, extremely dirty. It is a job best left to a professional.

The sound of my phonograph is terrible. Can I get a new reproducer for it?
Yes, but why not just rebuild yours? For a fraction of the cost of a
replacement, yours can probably be put into top-notch order by a
professional.

*Why do I have to change the needle in my 78 RPM disk phonograph at
least every other play? The point is still sharp.*
The needle is actually softer than the record and is designed to wear out.
Because of the heavy tracking weight of the tone arm, reproducer and
needle upon the record it was a choice of the needle or the record
receiving the wear. By changing the needle each play or no more than
every other play, wear to the record is minimized. It is not the point of
the record which receives the wear, but the taper. The wear is actually
microscopic: it cannot be seen without magnification, but the wear is
definitely there!

*When I bought my phonograph there were a lot of needles with it. Can I use
them?*
Unless you are <u>positive</u> that the needles are brand new, don't take a
chance. If they are lying loose in the machine, or are in a tray or needle
cup, chances are good that they are used needles. Brand new needles are
so inexpensive that it doesn't pay to take the chance of ruining a good
record.

I have heard that steel needles can be resharpened and reused. Is this so?
Not really. Steel needles are plated and once used the plating is worn
away. Sharpening with home tools will not give a smooth enough finish to
use on a record. New needles are quite inexpensive and will give the best
service.

Can I rotate the needle to get additional plays from it?
Generally rotating the needle will only change the angle of the wear to
the record and could actually accelerate the wear of the record. Put in
a new needle and feel secure.

My phonograph cabinet looks terrible. Should I strip it and refinish it?
Maybe, but first try cleaning it. Several commercial products are
available, like Murphy's Oil Soap. Follow the instructions carefully.
Pure turpentine is also a good and relatively mild cleaner. For any
cleaner, try it on an inconspicuous spot first. After cleaning, a good
lemon oil or boiled linseed oil/turpentine mixture will freshen the
cabinet. CAUTION: boiled linseed oil/turpentine is a flammable
mixture. Use in a well ventilated area and be sure to dispose of the
mixture properly <u>immediately</u> after using. Wash the rags **at once** in
warm, soapy water, and let dry in the fresh air.

Will a reproduction horn or other reproduction parts hurt the value of my phonograph?
Probably not, but much depends on the type of phonograph and which parts are replaced with reproductions. A phonograph with a reproduction horn is more complete and so more desirable than one with no horn at all. The same goes for the crank, the reproducer or the lid. You are better off having a reproduction than nothing at all. And it follows that the better the quality of the reproduction, the less it will hurt the value.

Will it hurt the value of my phonograph to refinish the cabinet and replate the metal parts?
Much depends on the type of phonograph and the condition of the cabinet. As a rule of thumb, the rarer the machine, the less desirable it is to refinish the cabinet. But if the cabinet and plating are so bad that they are an eyesore, then yes, refinish. Generally a phonograph in good original condition is worth more than excellent refinished condition.

If I refinish the cabinet, are decals available to complete the job?
In most cases yes they are. A new product is on the market today called 'Image Transfers'. These are not decals, and they give a much better finish than the traditional water decal. They are easy to use and virtually fool-proof. The two biggest advantages are that they don't leave a telltale bulge under the finished varnish coat as water transfers do, and second, they don't require drying time before varnishing or oiling.

What kind of finish should I use when I replate my metal parts?
If the parts were originally silver-colored, have them nickel plated, not chromed. If the color was originally gold, the plating was a gold plate. Today this is an expensive process. It sometimes pays just to have the parts brass-plated. But if it is a family heirloom or a rare machine, then spend the money and have the parts gold plated.

I cannot afford to buy another hand-wound phonograph. What can I collect that is related to these old phonographs that is reasonably priced?
The range of phonograph-related collectibles is almost endless. Once you start collecting you will discover many, many more ideas. A few suggestions are: unusual record labels; needle tins; antique newspaper and magazine articles; books on the history of phonographs (the earliest that we know of dates back to 1879); cylinder record boxes; Edison mementos; and the list can go on and on. . . .

I have a 'Music-Tone' brand (or other name) phonograph. Can you tell me more about it?

During the late 'teens and early 'twenties possibly hundreds of brands of floor and table model disk phonographs appeared on the market. Most were manufactured by small, independent companies. The cabinets were often contracted out and the motors, reproducers and tone arms were purchased from whoever had the best deal at the time. Some were quality machines, others were not. Little is documented about these smaller companies and their products.

How much should I insure my phonograph for?

There are several types of insurance coverage. For collectibles like wind-up phonographs you probably want replacement value. This means in case of a loss, the insurance company will pay what it would cost to replace it at today's prices. They may choose to replace it themselves for you. Contact your insurance agent for professional suggestions. A qualified antique's appraiser can put a replacement value on the phonograph. He will generally have to see the actual machine to do so. It cannot be done effectively with photographs or over the phone. Most appraisers charge a small fee for this service, often a percentage of the appraised valuation. It is suggested that you have clear photographs or a video recording of your collectibles. Store these outside the home with a trusted family member or in a bank safety deposit box.

I want to sell my phonograph. What's it worth?

It's not possible to give a blanket appraisal. Each machine varies in price depending on model, condition, accessories, horn, etc. And what is 'like new' condition to you might be only good condition to another collector. Each person has his own standards of judging. Many other factors go into selling a collectible like a phonograph. Geographic location, economic climate and time of year are just a few. You cannot put a value on nostalgia; someone else is not willing to pay a premium just because the phonograph was your great-grandfather's. Just because you saw one like yours in a local antique shop does not necessarily mean that you can get the same price. First of all, the antique shop may be way overpriced; secondly, most retailers offer service to their customers; some even extend some type of warranty. They also have to pay the expenses of having a store. That's part of their markup. Selling private party-to-private party is generally much less expensive than retail. Finally if you are selling to an antique dealer, don't expect him to pay you retail. He has to make a profit when he resells it. Often the small fee charged by an appraiser will be justified by a realistic selling price. A good local appraiser can also give you hints of where to best sell your collectible.

TROUBLE-SHOOTING

Just after the turn of the century, machines began to play a larger part in everyday life. A few years earlier virtually all chores in the home were done by hand, but technical developments made machines like washing machines and clothes manglers, bicycles and automobiles, electric fans and irons, and yes, even phonographs a common sight in the home. But people were not entirely trustful of these new devices, and repair facilities had to be relied on for even the most simple repairs. The machines were designed to be as service and attention free as possible.

The phonograph was one machine that was in constant use, so it had to be simple to use, reliable, and easily righted when something went wrong. Today we can thank the foresighted manufacturers for making these machines so reliable, but when they do go out of order there are some basic things to look for. Our repair suggestions are merely points to look for; they cannot be relied upon as absolute. Contact a professional repair service to repair your phonograph.

WHAT TO LOOK FOR WHEN...

PHONOGRAPH DOES NOT WORK AT ALL

Crank will turn but tension does not build up: mainspring(s) may be broken or disconnected. Contact a reliable repair professional. Often more than one spring will be broken. Broken mainsprings should be replaced, not repaired, riveted or welded.

Crank turns, tension begins to build up and then relaxes: Probably a broken mainspring. Contact a reliable repair professional.

Crank turns, tension builds up but motor does not run: brake not releasing, old grease binding in spring barrels or often a foreign object wedged between gears. Do not try to release mainspring or remove object. Contact a repairman.

Tension begins to build up but something in motor 'whirs', but turntable or record mandrel does not turn: broken, loose or stripped gear. Contact a repairman.

MOTOR WORKS BUT PHONOGRAPH DOES NOT OPERATE

Tension builds up, motor runs, but turntable or mandrel does not turn: broken, loose or missing belt (replace with correct belt); frozen or swelled mandrel bushing (on Edison cylinder phonographs - STANDARD, HOME, TRIUMPH generally models 'D' and later): contact a professional repairman to remove old bushing and replace.

Motor runs but tone arm (on disk phonographs) or carriage (on cylinder phonographs) does not move: On disk phonographs (most generally Columbia Grafonolas and similar) - pot metal tone arm frozen or swelled: <u>do not</u> force tone arm. Pot metal will break and repairs are difficult and often expensive. Contact a repairman. On cylinder phonographs (Edison) carriage frozen to carriage rod or half nut not engaging. Contact a professional to free carriage and replace or adjust half-nut. (Columbia) Pot metal carriage frozen or swelled. Contact a repairman.

MECHANISM OPERATES BUT PHONOGRAPH DOES NOT PLAY

Cylinder Phonographs: Is correct reproducer in place? Does reproducer have a stylus? Is reproducer tracking on the record? Have a professional repairman evaluate the reproducer.

Disk Phonograph: is there a needle in reproducer? Is diaphragm of reproducer broken or loose? Have a professional repairman evaluate the reproducer.

PHONOGRAPH PLAYS BUT SOUND IS NOT GOOD

Cylinder phonograph:

Check reproducer, horn and all connectors for physical blockages. Over the years paper, rags, insect's nests and other foreign objects might have blocked sound-flow. Remove blockage. Be sure rubber reproducer-to-horn connector is not crimped.

Reproducer probably needs work. A new stylus, linkage, gaskets or diaphragm can make it sound like new again. Have a professional repairman evaluate the reproducer.

Are you using the correct type of cylinder, correct 2 minute or four-minute with correct gearing and reproducer? Check type of record.

Is the record one of good quality? A bad record will <u>never</u> give good sound.

A wobbling or warbling sound is often caused by a loose belt. Replace belt with a new one.

Disk phonograph:

Check for obstructions in tonearm and in internal horn. Rags, insect's nests or other foreign matter could block the sound.

Reproducer probably needs work. New gaskets or diaphragm can make it sound like new again. Have a professional repairman evaluate the reproducer.

Are you using a <u>new</u> needle at least every other play? A used needle will give poor sound and ruin the record.

Are you using the correct type of record? Hand-cranked phonographs will <u>not</u> play modern 45 RPM or 33 1/3 RPM Long Plays. Even late (1950's-60's) 78's are often too soft to be played on an early phonograph.

PHONOGRAPH PLAYS BUT NOT WELL

A thumping or chattering sound occurs irregularly when phonograph is playing or being wound: grease in spring barrel is dry and must be replaced. Contact a repair professional.

Phonograph does not play all the way through a record: motor in need of lubrication or mainspring may be broken. Contact a phonograph repairman.

Strange mechanical noises, grinding or "knocking" noises. Could indicate serious trouble. Contact a phonograph repairman.

A phonograph found in an attic, barn or basement that hasn't been used for years <u>must</u> have fresh lubrication. Lack of lubrication is one of the greatest causes of machine failure. If your phonograph has been in the family for years and years, and you plan on keeping it and passing it on to your children, a good professional cleaning, lubrication and adjustments are recommended. Normal household oils and lubricants or automotive grease are not recommended for antique phonographs. For parts and repair services contact *Yesterday Once Again*. Their address and telephone number may be found at the end of the book.

BIBLIOGRAPHY

Generally a bibliography is listed alphabetically by author and title. Because of the special nature of this book, the bibliography is arranged by subject matter. In that way books of a specific nature can be easily identified and their importance to you pinpointed.

Many additional books on the subject of the phonograph exist; it is impossible to list them all. The books following are those which are most likely to be available to you. A large public, private or college library, and many natural science or special interest museums might also have a reference library for your use.

IDENTIFYING PHONOGRAPHS:

Baumbach, Robert Look for the Dog, 1994: identification of Eldridge Johnson, Victor and Victrola talking machines

Baumbach, Robert Columbia Phonograph Companion, Volume II - The Disc Graphophone and the Grafonola, 1996: an illustrated volume, invaluable in identifying the many Columbia disc machines

Baumbach, Robert Columbia Phonograph Companion, Volume I - The Cylinder Graphophone: anticipated publication late 1998

Bergonzi, Benet Old Gramophones, 1991: a history and identification of phonographs and gramophones (primarily European)

Contini, M. Fonografi E Grammofoni/Phonographs and Gramophones, 1987: over 100 individual models illustrated in full color. With text in both Italian and English.

Fabrizio, Timothy C. & Paul, George F. The Talking Machine - An Illustrated Compendium 1877-1929, 1997: an extremely fine text illustrated with 550 color photographs

Frow, George The Edison Cylinder Phonograph Companion, 2nd Edition 1994: identification of cylinder phonographs and reproducers manufactured by the Edison Phonograph Company

Frow, George The Edison Disc Phonographs and the Diamond Discs, 1982: identification of Edison disc phonographs

Jewell, Brian Veteran Talking Machines 1977: hundreds of phonographs pictured with brief description (primarily European)(out of print but still occasionally available)

Marty, Daniel The Illustrated History of Phonographs 1979: excellent photographs of U.S. and European phonographs (out of print)

Proudfoot, Christopher Collecting Phonographs and Gramophones 1980: many color illustrations of both U.S. and European phonographs (out of print)

IDENTIFYING RECORDS:

Barr, Stephen The Almost Complete Guide to 78 RPM Records II 1992: history and dating information of 78 records from late 1890's to 1942

Copeland, Peter Sound Recordings 1991: history and guide to recording techniques and (primarily) European recordings

Kinkle, Roger The Complete Encyclopedia of Popular Music and Jazz 1900-1950 four-volumes 1974: listing by title and artist of 78 RPMs in popular & jazz (out of print)

Koenigsberg, Allen Edison Cylinder Records 1889 - 1912 1987: Edison cylinders listed by cylinder number, artist, title

Rust, Brian Discography of Historical Records on Cylinders and 78s 1979: listing of historical and spoken word records (out of print)

Rust, Brian Jazz Records 1897-1942 2 Volumes 1978: listing by artists in jazz field on 78 RPMs (out of print)

Rust, Brian The American Dance Band Discography 1917-1942 2 volumes 1975: listing by title and artist of dance bands (out of print)

Rust, Brian The American Record Label Book 1984: hundreds of 78 labels pictured and discussed as to their background and history

Rust, Brian The Complete Entertainment Discography from the mid-1890's to 1942 1973: listing of major artists and their work by label and number (out of print)

Rust, Brian The Victor Master Book Vol. 2 1970: 78's on Victor listed by artist, title and label/matrix number. Volume 1, 3 not printed. (out of print)

Sherman, Michael The Paper Dog 1987: guide to 78 RPM Victor labels

Smart, James R. ed. The Edison Band - A Discography 1970: listing of recordings (cylinder & disk) made by Sousa's band (out of print)

Soderbergh, Peter Olde Records Price Guide 1979: prices for 78 RPM records 1900-1947 (newer editions are probably available)

Wile, Raymond R. & Dethlefson, Ronald, ed. Edison Disc Artists and Records 1910-1929 1990: a compilation of Edison Diamond Discs, Edison long play records and Edison 78 Rpm records

[Contact Nauck's Vintage Records for additional label and artist discographies: 6323 Inway Dr., Spring TX 77389]

GENERAL HISTORY OF THE PHONOGRAPH

Barnum, Fred His Master's Voice in America 1991: history of the Victor Talking Machine Company (and its predecessors and successors) in Camden, New Jersey, from the 1890's up to 1990 (General Electric). (Limited edition/out of print)

Frow, George The Edison Cylinder Phonograph Companion, 2nd Edition 1994: identification of cylinder phonographs and reproducers manufactured by the Edison Phonograph Company

Frow, George The Edison Disc Phonographs and the Diamond Discs, 1982: history and identification of Edison disc phonographs

Gelatt, Roland The Fabulous Phonograph 1955: the story of the Gramophone from tinfoil to high fidelity (out of print)

Koenigsberg, Allen ed. <u>The Patent History of the Phonograph 1877-1912</u> 1990: over 2,000 phonograph-related patents listed and explained

Miller, Russell & Roger Boar <u>The Incredible Music Machine 100 Glorious Years</u> 1982: the history of the talking machine, the artists and the recording companies (out of print)

Welch, Walter & Burt, Leah B.S. <u>From Tin Foil To Stereo</u> 2nd edition 1994: the evolution of the phonograph. Extensive details about people, places, events leading to the stereo phonograph as we know it. Primarily concerned with the early years

PHONOGRAPH OPERATION, CARE & MAINTENANCE

Reiss, Eric <u>The Compleat Talking Machine</u> 2nd edition 1996: the care, maintenance, repair and cosmetics of phonographs; with added identification and price guide

Phonograph Owner's Manuals, reprinted by *Yesterday Once Again*: copies of many of the original owners manuals supplied with the phonographs when new.

THE PEOPLE WHO MADE IT HAPPEN

Thomas Alva Edison

Many, many fine books have been written about Edison over the years. As we go to press several new works are in progress: the continuing Edison paper project and at least two new biographies. The works listed below might not always be the easiest to find, but they are among the best. . . .

Baldwin, Neil <u>Edison - Inventing the Century</u> 1995: a very fine biography of Edison from a different perspective

Jehl, Francis <u>Menlo Park Reminiscences</u> three volumes 1932: a chronological history of Edison's inventions by one of his co-workers since the 1870's (out of print)

Josephson, Matthew Edison 1959: one of the finest contemporary biographies of the man and his life (available in paperback)

Pretzer, William ed. Working at Inventing - Thomas Edison and the Menlo Park Experience 1989: several essays on Edison by noted authors

Wachhorst, Wyn Thomas Alva Edison - An American Myth 1981: another excellent biography (out of print)

Chichester Bell & Charles Sumner Tainter

Welch, Walter & Burt, Leah B.S. From Tin Foil To Stereo 2nd edition 1994: the evolution of the phonograph. Extensive details about people, places, events leading to the stereo phonograph as we know it. Primarily concerned with the early years

Emile Berliner

Barnum, Fred His Master's Voice in America 1991: history of the Victor Talking Machine Company (and its predecessors and successors) in Camden, New Jersey, from the 1890's up to 1990 (General Electric). (Limited edition/out of print)

Gelatt, Roland The Fabulous Phonograph 1955: the story of the Gramophone from tinfoil to high fidelity (out of print)

Wile, Frederic William Emile Berliner - Maker of the Microphone 1926: a fine biography of the inventor of the disk record & more (out of print)

Eldridge R. Johnson

Barnum, Fred His Master's Voice in America 1991: history of the Victor Talking Machine Company (and its predecessors and successors) in Camden, New Jersey, from the 1890's up to 1990 (General Electric). (Limited edition/out of print)

Johnson, E.R. Fenimore His Master's Voice Was Eldridge R. Johnson 1974: a biography of Johnson by his son. Very interesting details of his life (out of print but still available)

Johnson, Mrs. Eldridge Reeves <u>Eldridge Reeves Johnson -</u> <u>Industrial Pioneer</u> 1951: pamphlet biography compiled by his wife (out of print but still occasionally available)

Welch, Walter & Burt, Leah B.S. <u>From Tin Foil To Stereo</u> 2nd edition 1994: the evolution of the phonograph. Extensive details about people, places, events leading to the stereo phonograph as we know it. Primarily concerned with the early years

NIPPER

Barnum, Fred <u>His Master's Voice in America</u> 1991: history of the Victor Talking Machine Company (and its predecessors and successors) in Camden, New Jersey, from the 1890's up to 1990 (General Electric). (Limited edition/out of print)

Edge, Ruth & Petts, Leonard <u>The Collectors Guide to 'His Master's</u> <u>Voice' Nipper Souvenirs</u> 2nd edition 1997: a detailed listing of 'Nipper' collectibles compiled from both public and private collections throughout the U.S.A., Europe and Asia

Petts, Leonard <u>The Story of 'Nipper' and the 'His Master's Voice'</u> <u>Picture</u> 1983: a very detailed history of the dog, the painter and the famous painting of 'His Master's Voice'

SEVERAL UNITED STATES MUSEUMS WITH COLLECTIONS RELATING TO PHONOGRAPHS

Edison National Historic Site, Main Street & Lakeside Ave, West Orange, NJ 07052 (201) 736-5050 (Edison's main laboratory facility)

Henry Ford Museum and Greenfield Village, Ann Arbor, MI 48121 (313) 271-1620 (Menlo Park restoration)

Edison Birthplace Museum, 9 Edison Drive, Milan, OH 44846 (419) 499-2135 (collection of Edison and phonograph artifacts)

Edison's Winter Home, 2350 McGregor Blvd, Fort Myers, FL 33901 (813) 334-3614 (Edison & phonograph memorabilia)

Edison Plaza Museum, Pine Street, Beaumont, TX 77704 (409) 839-3089 (Edison & phonograph artifacts)

Edison Memorial Tower and Museum, 37 Christie Street, Edison, NJ 08820 (201) 549-3299 (Edison artifacts)

Bellem's Cars and Music of Yesterday, 5500 N Tamiami Trail, Sarasota, FL 33580 (813) 355-6228

Johnson Victrola Museum, Delaware State Museum, 316 S. Governor's Ave, Dover, DE 19901 (302) 736-4266 (Victor & Victrola phonographs, records and artifacts)

Smithsonian Museum, Division of Science and Technology, Washington, DC 20560 (202) 357-1840 (rare and unusual phonographs)

Seven Acres Museum, 8512 S Union Road, Union, IL 60180 (815) 923-2214

Library of Congress, Collection of Recorded Sound, Washington, DC 20540 (202) 707-5522 (recordings, music and artifacts)

PHONOGRAPH REPAIRS, PARTS, BOOKS & SUPPLIES

Yesterday Once Again, P.O. Box 6773, Dept. BK, Huntington Beach, CA 92615 (714) 963-2474: Complete repair facilities and machine shop; original and reproduction parts, mainsprings, reproducers. Books, manuals, needles, record sleeves, Nipper, accessories. Free catalog.

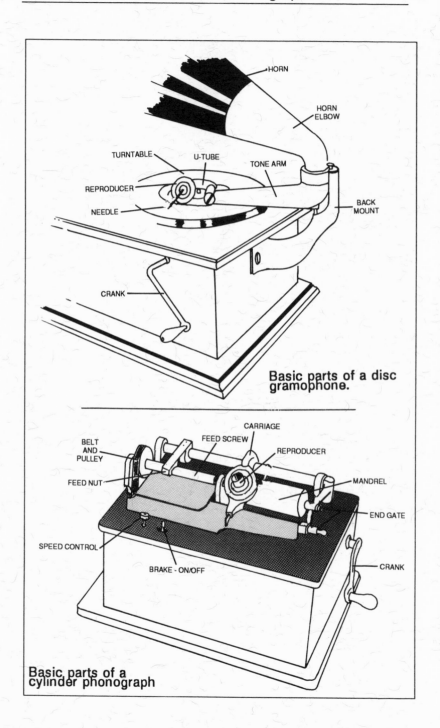

Basic parts of a disc gramophone.

Basic parts of a cylinder phonograph

Glossary

AMBEROLA Edison's inside-horn cylinder phonograph.

BACK MOUNT A cast-iron bracket which bolts onto the back of the cabinet. It supports both the tone arm and the horn assembly.

BELT (Or drive-belt) A leather belt which connects a pulley on the lower motor with the mandrel on the upper assembly of the cylinder phonograph.

CARRIAGE a metal arm which holds both the reproducer and the feed nut. It is aligned so that the needle moves across the surface of the cylinder record and is moved by the feed screw.

CELLULOID an early plastic-type material used for durable cylinder record surfaces.

COIN-IN-THE-SLOT PHONOGRAPH a specially converted phonograph with a 'lock' to prevent unauthorized playing of the record. The 'key' to this lock was a coin, most often a nickel. This released the mechanism and allowed the listener to hear the selection.

CRANK the winding handle which is turned 'clockwise' to wind the spring(s) in the motor.

CYLINDER a tube-shaped record designed to be played on a cylinder phonograph. The grooves are on the outside of the tube, playing from left to right.

DIAMOND DISC (1) a special 1/4" thick record manufactured by Edison with his vertical recording method. Designed to be played only on Edison's Diamond Disc phonographs. (2) Reference to the special disc phonographs made by Edison to play his vertically recorded records.

DIAPHRAGM a thin piece of material, most often mica, metal or specially-treated paper which vibrates when the impressions of sound from the record are transferred to it. It converts the sound images into audible sounds and begins to amplify them.

DISK or **DISC** a flat plate-shaped record with grooves either on one side or both. The grooves are concentric, in most cases beginning at the outside and spiraling in toward the center. (Although either spelling is correct, disk seems to be the preferred; Edison used the spelling disc for his phonographs and so I have tried to follow suit.)

END GATE A swinging bracket found on many of the earlier cylinder phonographs which supported the right end of the mandrel shaft.

FEED NUT or **HALF NUT** a threaded piece of steel with the threads and contour matching the feed screw. The feed nut is attached to the carriage and engages the feed screw when the machine is playing.

FEED SCREW a helical threaded rod which is connected to the cylinder phonograph motor and moves the carriage from left to right.

FOUR-MINUTE CYLINDER a tube-shaped record, generally the same physical size as a two-minute record, but with twice as many grooves and approximately four minutes playing time.

FRONT-MOUNT PHONOGRAPH an early style disk phonograph with the horn supported from the front of the machine in a wire cradle. The reproducer attached directly to end of the horn often through a metal or leather elbow.

GASKETS thin rings of cushioning material surrounding the edges of the diaphragm within the reproducer to keep the diaphragm flexing properly.

GRAMOPHONE (1) The disk-style talking machine invented by Emile Berliner, (2) European usage: a generic-type reference to any disk record player.

GRAFONOLA the inside-horn version of the Columbia talking machine.

GRAPHOPHONE The name given to the talking machine developed by Bell and Tainter - later the Columbia.

HEAD see Reproducer

HILL & DALE see vertical recording

HORN generally a cone shaped tube (ranging anywhere from 10" up to over four feet) designed to amplify the sound from the reproducer. In recording, the horn compresses the sound for a stronger impression.

HORN CRANE an array of metal rods designed to attach to the cylinder phonograph and support the weight of the larger horns.

HORN ELBOW a connector which fits between the horn and the back mount bracket on a rear-mount disk phonograph.

HORN FLOOR STAND a free-standing metal rack with a curved top section to support the larger horns.

LATERAL RECORDING a recording system where the sound is placed on the 'side' of the groove. The needle will move back and forth in 'feeling' the sound and retrieving it from the record.

LISTENING TUBES similar to a doctor's stethoscope, connected directly to the reproducer and fitting into the listener's ears to hear the sound.

MAINSPRING a flat strip of spring-steel, coiled into (generally) a metal barrel. May be used singly or in multiples depending on phonograph. When wound with a crank, it supplies the power to operate the phonograph motor. CAUTION: mainspring repairs or examination should not be attempted by an amateur. Serious injuries could occur.

MANDREL a tapered metal tube on a cylinder phonograph which is designed to hold the cylinder record while it is playing.

NEEDLE see stylus

PHONOGRAPH (1) The talking machine invented by Thomas Alva Edison in 1877, (2) a generic-type reference to any record player.

POT METAL an inexpensive method of cast alloy of metals. Often used for parts which would have been too expensive to machine or extrude. Pot metal (also sometimes called white metal) turns very brittle with age and loses most of its strength. Over the years pot metal parts might swell, crack, chip or shatter. Repairs are generally not successful.

REAR-MOUNT PHONOGRAPH the later style disk phonograph where a cast-iron support bolted onto the back of the phonograph. This metal support held a tone arm and the reproducer and also the horn assembly.

RECORD any medium (wax, shellac, celluloid, etc.) or format (disk, cylinder) designed to have sound registered on it, and with the ability to play-back the same sounds.

REPEATER a device attached to a phonograph allowing either the entire record to be replayed automatically or for a single groove to be replayed.

REPRODUCER also known as a soundbox or a head; the sound is picked up by the needle or stylus of the reproducer from the record, converted by the diaphragm within the reproducer into audible sounds and amplified.

SHAVER a device, either part of the phonograph itself, or as a separate machine entirely, designed to scrape (shave) the old recorded surface off brown wax records, leaving a clean, smooth surface for a new recording.

SPEED CONTROL A device attached to the motor which regulates the speed (number of revolutions per minute or RPMs) of the record. It can be found in different locations on various phonographs. The correct speeds for most records are: disk records - 78 RPM; Diamond Disc records - 80 RPM; most cylinder records 160 RPM; (earlier cylinder records were 120 RPM and earlier disks varied considerably in speed).

SOUNDBOX see Reproducer

SPINDLE the shaft in the center of the turntable. It fits through the hole in the record.

STYLUS the point or needle which contacts the record and rubs against the sound impressions, transferring them from the record to the diaphragm. A stylus is generally a jewel; a needle is often steel.

TIN FOIL a thin foil of tin and lead, similar in appearance to aluminum foil, used as the first material for recording.

TONE ARM a tube connecting the reproducer to the horn assembly, most often on a disk phonograph.

TURNTABLE a metal platter onto which the disk record fits. The spindle shaft of the motor fits through the center hole in the turntable, and the power from the motor causes the turntable to revolve.

TWO-MINUTE CYLINDER a tube-shaped record of approximately two minutes playing time.

VERTICAL RECORDING a recording system where the sound is placed on the 'bottom' of the groove. The needle will move up and down in 'feeling' the sound and retrieving it from the record. Also known as Hill & Dale.

VICTROLA (1) the designation of all talking machines manufactured by Victor with the horn concealed <u>within</u> the cabinet, (2) a generic-type reference to any record player (incorrect).

HAND-CRANKED PHONOGRAPHS - *It All Started With Edison* is Neil Maken's first book. Maken's love for the hand-cranked phonograph began with a Victrola which he purchased in a thrift shop, took home, repaired, and got to play. He became fascinated with the old phonograph, started reading and studying the various types of talking machines, bought another, and then another, and then.... His hobby grew into a part-time business of repairs and sales, and eventually grew to the point that he was forced to resign his 9-5 sales management job and make phonographs a full-time responsibility.

Maken's writings have been published internationally in magazines, newspapers and newsletters. He works closely with several U.S., Asian and European museums which specialize in phonographs and Thomas Edison. He has been seen on the History Channel's biography of Thomas Edison and worked closely with the production staff of that filmed biography.

Neil Maken was raised on Long Island, NY, and graduated from Long Island University. He now lives in Southern California with his wife, Carole, his daughter, Tracey and their pug dog, Cleo.